T0091601

Mathematical Theory of
Liquid Interfaces
Liquid Layers, Capillary Interfaces,
Floating Drops and Particles

Contents

Values for the surface tension and boundary contact angle concerning different liquids and materials one finds in [Landolt-Börnstein (1956)].

Pierre Simon Laplace [Laplace (1806)] derived the equilibrium condition

$$p_1 - p_2 = 2\sigma H$$

for the interface, where p_1 denotes the pressure in the liquid, p_2 is the pressure outside the liquid, σ is the surface tension and H denotes the mean curvature of the interface. The notation of the mean curvature was introduced by Thomas Young [Young (1805)].

1.1 Mean curvature, Gauss curvature

For two vectors $a = (a_1, a_2, a_3)$, $b = (b_1, b_2, b_3)$ in $\mathbb{R}^3$, we denote the *inner product* or *scalar product* by $\langle a, b \rangle$ or $a \cdot b$, which is defined by

$$\langle a, b \rangle = \sum_{k=1}^{3} a_k b_k.$$

The *exterior product* $a \wedge b$ or $a \times b$, is given by

$$a \wedge b = (a_2 b_3 - a_3 b_3, -a_1 b_3 + a_3 b_1, a_1 b_2 - a_2 b_1),$$

which can be written formally as

$$a \wedge b = \det \begin{pmatrix} e_1 & e_2 & e_3 \\ a_1 & a_2 & a_3 \\ b_1 & b_2 & b_3 \end{pmatrix},$$

where e_1, e_2 and e_3 denote the standard basis in $\mathbb{R}^3$.

Assume that $x : u \in U \mapsto \mathbb{R}^3$, $u = (u^1, u^2)$ or $u = (\alpha, \beta)$, defines a regular surface $\mathcal{S}$, where $U \subset \mathbb{R}^2$ is a domain in $\mathbb{R}^2$. Regular means that all involved functions are sufficiently regular and that the vectors x_{u^1} and x_{u^2} are linearly independent. Concerning the definition and properties of curves and surfaces in $\mathbb{R}^n$ see for instance [Ossermann (1986)] and [Dierkes, Hildebrandt, Küster and Wohlrab (1992)]. The normal $N = N(u)$ on $\mathcal{S}$ is defined by

$$N = \frac{x_{u^1} \wedge x_{u^2}}{|x_{u^1} \wedge x_{u^2}|}.$$

The coefficients of the first fundamental form are given by

$$E = \langle x_\alpha, x_\alpha \rangle, \ F = \langle x_\alpha, x_\beta \rangle, \ G = \langle x_\beta, x_\beta \rangle,$$

and the coefficients of the second fundamental form are

$$l = -\langle N_\alpha, x_\alpha \rangle, \ m = -\langle N_\alpha, x_\beta \rangle = -\langle N_\beta, x_\alpha \rangle, \ n = -\langle N_\beta, x_\beta \rangle.$$

The mean curvature $H = H(u)$ is defined by

$$H = \frac{lG + nE - 2mF}{2(EG - F^2)},$$

and the Gauss curvature by

$$K = \frac{ln - m^2}{EG - F^2}.$$

Remark 1.1. In terms of the principle curvatures R_1, R_2, see [Blaschke (1921)], pp. 57-59, one has

$$H = \frac{1}{2}\left(\frac{1}{R_1} + \frac{1}{R_2}\right), \qquad K = \frac{1}{R_1 R_2}.$$

An important formula which we will use frequently is

$$\triangle x = 2HN, \tag{1.2}$$

where $\triangle$ is the Laplace-Beltrami operator on $\mathcal{S}$, see [Dierkes, Hildebrandt, Küster and Wohlrab (1992)], [Dierkes, Hildebrandt, and Sauvigny (2010)] p. 71, p. 72, resp., for a proof of this formula.

The following formula will be used concerning the question whether or not a given equilibrium surface defines a strong minimizer of the associated energy functional. Consider a family $\{\mathcal{F}_c\}$ of regular surfaces which can be described as level surfaces

$$\mathcal{F}_c = \{x \in \mathbb{R}^3 : s(x) = c\}.$$

Suppose that $\nabla s(x) \neq 0$, then

$$Q(x) := \frac{\nabla s(x)}{|\nabla s(x)|}$$

is orthogonal to all surfaces $\mathcal{F}_c$, see Fig. 1.4, and

$$\text{div } Q(x) = -2H(x), \tag{1.3}$$

where $H(x)$ is the mean curvature at x of the surface of the family which contains x. See [Dierkes, Hildebrandt, Küster and Wohlrab (1992)], p. 77, for a proof of this formula. Here $x = x(u)$ denotes a C^2-parametrization of $\mathcal{F}_c$ such that $N(u) = Q(x(u))$.

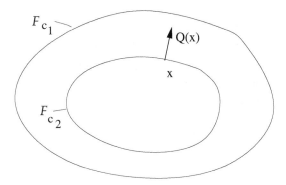

Fig. 1.4 Level curves

1.2 Liquid layers

Porous materials have a large amount of cavities, different in size and geometry. Such materials swell and shrink in dependence on air humidity. Here we consider an isolated cavity, see [Schiller, Mögel and Miersemann (2010)] for some cavities of special geometry.

Let $\Omega_s \in \mathbb{R}^3$ be a domain occupied by a homogeneous solid material. The question is whether or not liquid layers Ω_l on Ω_s are stable, where Ω_v is the domain filled with vapor and $\mathcal{S}$ is the capillary surface which is the interface between liquid and vapor, see Fig. 1.5.

It is assumed that the boundary of the solid material is completely covered with liquid.

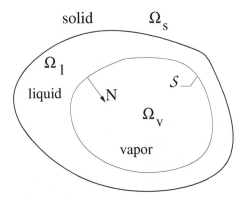

Fig. 1.5 Liquid layer in a pore

Let

$$\mathcal{E}(\mathcal{S}, \mu) = \sigma|\mathcal{S}| + w(\mathcal{S}) - \mu|\Omega_l(\mathcal{S})|$$

be the energy, the grand canonical potential, of the problem, where σ is the surface tension, μ a positive constant, $|\mathcal{S}|$ and $|\Omega_l(\mathcal{S})|$ denote the area resp. volume of $\mathcal{S}$ and $\Omega_l(\mathcal{S})$, and

$$w(\mathcal{S}) = -\int_{\Omega_v(\mathcal{S})} F(x)\, dx ,$$

denotes the disjoining pressure potential, where

$$F(x) = c\int_{\Omega_s} \frac{dy}{|x-y|^p} .$$

Here c is a *negative* constant, $p > 4$ a positive constant ($p = 6$ for nitrogen) and $x \in \mathbb{R}^3 \setminus \overline{\Omega}_s$. The closure $\overline{\Omega}_s$ of Ω_s is the union of Ω_s with its boundary $\partial\Omega_s$. The disjoining pressure potential prevents the interface $\mathcal{S}$ to meet the container wall.

Suppose that $\mathcal{S}_0$ defines a local minimum of the energy functional, then

$$-2\sigma H + F - \mu = 0 \ \text{ on } \mathcal{S}_0 ,$$

where H is the mean curvature of the interface $\mathcal{S}_0$.

1.3 Capillary interfaces

Consider a container partially filled with a bounded amount of liquid, see Fig. 1.6. Suppose that the associate energy functional is given by

$$\mathcal{E}(\mathcal{S}) = \sigma|\mathcal{S}| - \sigma\beta|\mathcal{W}(\mathcal{S})| + \int_{\Omega_l(\mathcal{S})} F(x)\, dx,$$

where $F = Y\rho$, and
Y potential energy per unit mass, for example $Y = gx_3$, $g = const. \geq 0$,
ρ local density of the liquid,
σ surface tension, $\sigma = const. > 0$,
β, $-1 \leq \beta \leq 1$, denotes here the constant (relative) adhesion coefficient between the fluid and the container wall (sometimes β denotes also the second coordinate in the parameter couple $u = (\alpha, \beta)$),
$\mathcal{W}$ wetted part of the container wall,
Ω_l domain occupied by the liquid.
Additionally, we have for given volume V of the liquid the constraint

$$|\Omega_l(\mathcal{S})| = V.$$

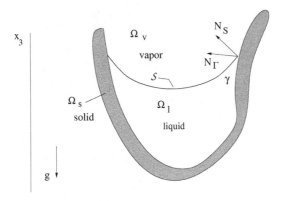

Fig. 1.6   Liquid in a container

It turns out that a minimizer $\mathcal{S}_0$ of the energy functional under the volume constraint satisfies, see [Finn (1986)], Chap. 1,

$$2\sigma H = \lambda + F \quad \text{on } \mathcal{S}_0,$$
$$\cos\gamma = \beta \quad \text{on } \partial\mathcal{S}_0,$$

where H is the mean curvature of $\mathcal{S}_0$, λ a constant Lagrange parameter and γ is the angle between the surface $\mathcal{S}_0$ and the container wall at $\partial\mathcal{S}_0$.

Remark 1.2. The angle between two surfaces at a point of intersection is by definition the angle between the associated normals at this point.

Remark 1.3. The term $-\sigma\beta|\mathcal{W}|$ in the above energy functional is called *wetting energy*.

Remark 1.4. The above boundary condition $\cos\gamma = \beta$ shows that the contact angle γ is constant along the boundary $\partial\mathcal{S}_0$.

In the case that the capillary surface $\mathcal{S}$ is a graph over $\Omega \subset \mathbb{R}^2$, i. e., $x_3 = u(x_1, x_2)$ defines the capillary surface, and if the container wall is the cylinder surface $\partial\Omega \times \mathbb{R}^+$, then the related boundary value problem is

$$\operatorname{div} Tu = \frac{g\rho}{\sigma}u + \frac{\lambda}{\sigma} \quad \text{in } \Omega,$$
$$\nu \cdot Tu = \cos\gamma \quad \text{on } \partial\Omega,$$

where ν is the exterior unit normal at $\partial\Omega$, and

$$Tu := \frac{\nabla u}{\sqrt{1 + |\nabla u|^2}},$$

div Tu is twice the mean curvature of the surface defined by $x_3 = u(x_1, x_2)$, see an exercise.

The above problem describes the ascent of a liquid, water for example, in a vertical cylinder with constant cross section Ω. It is assumed that gravity is directed downwards in the direction of the negative x_3-axis. A liquid can rise along a vertical narrow wedge, see Fig. 1.7, which is a consequence of the strong non-linearity of the underlying equations.

Fig. 1.7   Ascent in the edge of a narrow wedge

Another interesting question is whether or not liquid rests, a drop for example, on an inclined plane or on a ring, see Fig. 1.8.

Fig. 1.8   Drop on a ring

1.4 Floating drops

For a floating drop, for example a drop of balsamico on olive oil, see Fig. 1.9, Franz Neumann [Neumann (1894)] derived equilibrium conditions which satisfy the involved surfaces, see also [Slobozhanin (1986)]. Initial results concerning a careful mathematical analysis of the drop problem were achieved in [Elcrat, Neel and Siegel (2004)].

Fig. 1.9   Drop of balsamico on olive oil

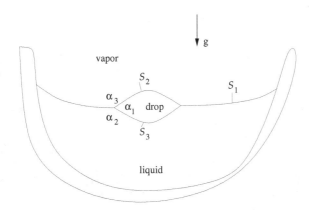

Fig. 1.10   Floating drop, notations

In Fig. 1.10 we follow the notations of [Slobozhanin (1986)]. One of the equilibrium conditions is the angle condition

$$\cos \alpha_i = \frac{\sigma_i^2 - \sigma_j^2 - \sigma_k^2}{2\sigma_j \sigma_k},$$

where $i \neq j$, $i \neq k$, $j \neq k$, and σ_k are the surface tensions of the interfaces $\mathcal{S}_k$, see Fig. 1.10. The above formula shows that the angles α_k are constant along the edge of the drop.

1.5   Floating particles

Consider a particle on a surface of a liquid which fills partially a container. It is assumed that gravity g is directed downwards in the direction of the negative x_3-axis. Bodies of density exceeding that of the liquid can float on a liquid. Examples are paper clips, see Fig. 1.11, needles or razor blades. Even for small particles or if the gravity is small, the surface tension is the dominating force.

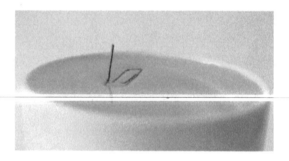

Fig. 1.11   Floating paperclip

Some methods of measurement of surface tension are based on the equilibrium conditions for floating bodies.

Fig. 1.12   Floating balls

Floating particles can attract or repel each other, can move to the boundary of the container or can move away from the boundary. Different contact angles and positions of the particle determine this behavior.

For example, a ping-pong ball can stay in the middle of a glass filled with water a little bit over the height of the glass, and the ball moves to the boundary if there is less water in the glass, see Fig. 1.12.

Fig. 1.13   Floating thumbtacks

In the case of a floating thumbtack one observes the opposite behavior, see Fig. 1.13.

1.6   Wetting barriers

Liquid can hang on the edge of a container, see Fig. 1.14. Here the capillary

Fig. 1.14   Liquid hangs at the upper edge

surface $\mathcal{S}$ satisfies a variational inequality at $\partial\mathcal{S}$ where $\mathcal{S}$ meets the container wall along an edge, see [Miersemann (2002)]. Such an edge is called *wetting barrier*. The following pictures show more examples of wetting barriers

Fig. 1.15   A drop with a hole

Fig. 1.16   Liquid hangs partially at the upper edge

Fig. 1.17   Tubes with constant and variable cross section

Another kind of a wetting barrier is a curve on the container wall which separates domains with different adhesion coefficients. Assume that the adhesion coefficient on the wetted part is different from the adhesion coefficient on the non-wetted part of the container wall then the contact angle of a liquid in equilibrium is in general not constant along the boundary of the liquid.

1.7   Contact angle

The theory of [Gauss (1839)], see Chap. 3, says that the contact angle is a constant which contradicts experiments. Consider a drop sitting on an inclined plane in the presence of a gravitational field directed downwards, see Fig. 1.18. It was shown in [Shinbrot (1985)], see also [Finn (1986)], Chap. 8, that the contact angle cannot be a constant. This contradiction between the theory of Gauss and the experiment with a drop on an inclined plane suggests the conjecture that the adhesion coefficients β_w and β_{nw} on the wetted part and on the non-wetted part are probably different from each

(8) Find an energy functional $\mathcal{E}$ such that its extremals satisfy

$$\operatorname{div} Tu = \frac{g\rho}{\sigma}u + \frac{\lambda}{\sigma} \text{ in } \Omega,$$
$$\nu \cdot Tu = \cos\gamma \text{ on } \partial\Omega.$$

Chapter 2

Liquid layers

The classical macroscopic theory of capillarity is usually applied to describe the location and the behavior of the fluid in a container. The interface between the liquid and the vapor makes a constant contact angle with the container wall if the container is made from homogeneous material, according the general theory of [Gauss (1839)]. The associated strongly nonlinear boundary value problem which describes the interface arose from studies of [Young (1805); Laplace (1806); Gauss (1839)].

A refined mesoscopic description incorporates the disjoining pressure which results from the long range forces between the fluid and the solid substrates. The theory of disjoining pressure of liquid layers was initiated by [Derjaguin (1939, 1940, 1946)]. In [Schiller, Wahab and Mögel (2004)], in particular, a single wedge is considered as a special geometry of solid material.

Let $\Omega_s \in \mathbb{R}^3$ be a domain occupied by homogeneous solid material, Ω_l the liquid layer, Ω_v the domain filled with vapor and $\mathcal{S}$ is the capillary surface which is the interface between liquid and vapor, see Fig. 1.5. Let

$$\mathcal{E}(\mathcal{S}, \mu) = \sigma|\mathcal{S}| + w(\mathcal{S}) - \mu|\Omega_l(\mathcal{S})| \qquad (2.1)$$

be the energy (grand canonical potential) of the problem, where σ denotes the surface tension and $|\mathcal{S}|$, $|\Omega_l(\mathcal{S})|$ the area resp. volume of $\mathcal{S}$ and $\Omega_l(\mathcal{S})$,

$$w(\mathcal{S}) = -\int_{\Omega_v(\mathcal{S})} F(x) \, dx \;, \qquad (2.2)$$

is the disjoining pressure potential, where

$$F(x) = c \int_{\Omega_s} \frac{dy}{|x - y|^p} \; . \qquad (2.3)$$

Here c is a *negative* constant, $p > 4$ a positive constant ($p = 6$ for liquid nitrogen) and $x \in \mathbb{R}^3 \setminus \overline{\Omega}_s$, where $\overline{\Omega}_s$ is the closure of Ω_s, i. e., the union of Ω_S with its boundary $\partial\Omega_s$. Finally set

$$\mu = \rho k T \ln(X) ,$$

where
ρ density of the liquid,
k Boltzmann constant,
T absolute temperature,
X reduced (constant) vapor pressure, $0 < X < 1$.

More precisely, ρ is the difference between the densities of the liquid and the vapor phase. However, since in most practical cases the vapor density is rather small, ρ can be replaced by the density of the liquid phase.

The above negative constant is given by $c = \mathcal{H}/\pi^2$, where $\mathcal{H}$ is the Hamaker constant, see [Israelachvili (1995)], p. 177. For a liquid nitrogen film on quartz one has about $\mathcal{H} = -10^{-20}\,Nm$.

Ansatz (2.1) can be justified by using density functional theory for fluids in combination with the sharp-kink approximation, see [Bieker and Dietrich (1998)]. In our approach we neglect a small curvature correction of the surface tension. This correction is negligibly small if curvature radii are much larger than a molecular diameter 0.3 nm. Formulas and asymptotic expansions of $F(x)$ for plane, spherical, cylindrical and wedge geometries are calculated by [Philip (1977)].

Let $\mathcal{S}(\epsilon)$, $|\epsilon| < \epsilon_0$, be the family of comparison surfaces given by

$$z(u, \epsilon) = x(u) + \epsilon\xi(u)N(u),$$

where $x(u)$, $u \in U \subset \mathbb{R}^2$, $u = (u^1, u^2) = (\alpha, \beta)$, is a parameter representation of the surface $\mathcal{S}_0 = \mathcal{S}(0)$ such that the unit normal on $\mathcal{S}_0$

$$N = \frac{x_\alpha \wedge x_\beta}{|x_\alpha \wedge x_\beta|}$$

is *directed out* of the fluid into the vapor. It is assumed that $\mathcal{S}_0$ and the scalar function $\xi(u)$ are sufficiently regular.

In the case of capillary interfaces the boundary of the admissible comparison surfaces $\mathcal{S}(\epsilon)$ must lie on the boundary of the solid domain, see [Finn (1986)], p. 7, for the construction of such surfaces. In our choice of energy the liquid covers completely the solid surface since the function F defined by (2.3) becomes unbounded near the solid surface.

2.1 Governing formulas

A necessary condition such that $\mathcal{S}_0$ defines a local minimum of the energy functional $\mathcal{E}(\mathcal{S}, \mu_0)$ is

$$\left[\frac{d}{d\epsilon}\mathcal{E}(\mathcal{S}(\epsilon), \mu_0)\right]_{\epsilon=0} = 0 \, ,$$

which leads to the equation, see the corollaries to Lemmas 2.2–2.4 of the appendix to this chapter,

$$-2\sigma \int_{\mathcal{S}_0} H\xi \, dA + \int_{\mathcal{S}_0} F\xi \, dA - \mu \int_{\mathcal{S}_0} \xi \, dA = 0$$

for all ξ, where H is the mean curvature of $\mathcal{S}_0$, see Sec. 1.1. An interface $\mathcal{S}_0$ is said to be an equilibrium surface if this equation holds.

Theorem 2.1. *Suppose that $\mathcal{S}_0$ defines an equilibrium, then on $\mathcal{S}_0$*

$$- 2\sigma H + F - \mu = 0. \tag{2.4}$$

An existing equilibrium state $\mathcal{S}_0$ is said to be *stable* by definition if

$$\left[\frac{d^2}{d\epsilon^2}\mathcal{E}(\mathcal{S}(\epsilon), \mu_0)\right]_{\epsilon=0} > 0$$

for all ξ not identically zero. From the corollaries to the Lemmas 2.5–2.7 of the appendix to this chapter, integration by parts and by using the equilibrium condition (2.4) we find that

Theorem 2.2. *An equilibrium interface $\mathcal{S}_0$ is stable if*

$$\sigma \int_{\mathcal{S}_0} \left(|\nabla \xi|^2 - 2(2H^2 - K)\xi^2 + \langle \nabla F, N \rangle \, \xi^2\right) \, dA > 0 \tag{2.5}$$

for all $\xi \in W^{1,2}(\mathcal{S}_0) \setminus \{0\}$.

Here K is the Gauss curvature of the capillary surface $\mathcal{S}_0$, see Sec. 1.1. Concerning the definition of $\nabla \xi$ on $\mathcal{S}_0$ see for instance [Blaschke (1921)], p. 113. The Sobolev space $W^{1,2}(\mathcal{S}_0)$ may be replaced by the space $C^1(\mathcal{S}_0)$ of continuously differentiable functions. For the definition of Sobolev spaces see [Adams (1975)] for example. If $\mathcal{S}_0$ is unbounded, then we assume that ξ has compact support, i. e., $\xi \equiv 0$ on $\mathcal{S}_0 \setminus M$, where M is a compact set of $\mathcal{S}_0$.

Thus, since there is no side condition on ξ, $\mathcal{S}_0$ is stable if

$$- 2(2H^2 - K) + \frac{1}{\sigma}\langle \nabla F, N \rangle > 0 \quad \text{on } \mathcal{S}_0. \tag{2.6}$$

When the left hand side of (2.6) is constant on S_0, which is satisfied in all examples considered in the following, then (2.6) is also necessary for stability.

If inequality (2.5) is satisfied, then S_0 defines at least a weak local minimum of the energy functional (2.5), i. e., $\mathcal{E}(S_0, \mu_0) \leq \mathcal{E}(S, \mu_0)$ for all surfaces S in a C^1-neighborhood of S_0. This follows from the Taylor expansion of the associated energy functional, see [Miersemann (2002)], where capillary interfaces are considered.

2.2 Explicit solutions

In this section we consider examples of pores with simple geometry where we know the equilibrium state, i. e., surfaces S_0 which satisfy the necessary condition (2.4). Then we ask whether or not such a surface is stable in the sense that the condition (2.6) is satisfied. In general, for given solid surface the equilibrium states are not known explicitly.

2.2.1   Adsorption in slit cavities

Assume that two parallel plates have a distance $2d$ from each other, and the regions $y_3 > d$ and $y_3 < -d$ are occupied by the solid, see Fig. 2.1.

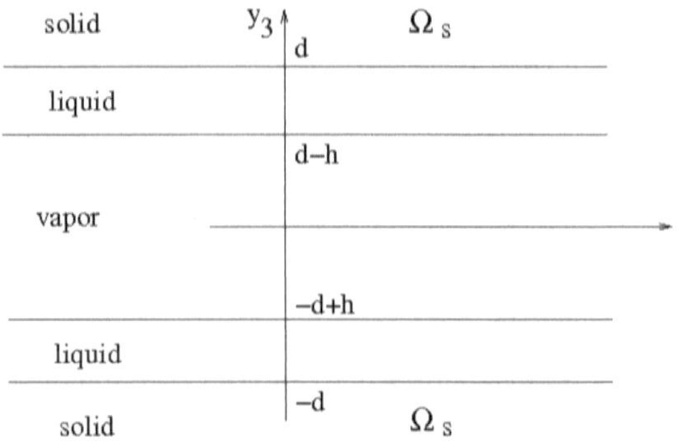

Fig. 2.1   Slit cavity

The potential (2.3) is given by

$$F(x) = c \int_{y_3 > d} \left(y_1^2 + y_2^2 + (x_3 - y_3)^2 \right)^{-p/2} dy$$

$$+ c \int_{y_3 < -d} \left(y_1^2 + y_2^2 + (x_3 - y_3)^2 \right)^{-p/2} dy \ .$$

The plane surfaces $\mathcal{S}_1 : y_3 = d - h$ and $\mathcal{S}_2 : y_3 = -d + h$ define stable equilibria if the equilibrium condition $F - \mu = 0$ on $\mathcal{S}_i$ as well as the stability condition $\partial F / \partial N \equiv \langle \nabla F, N \rangle > 0$ on $\mathcal{S}_i$ are satisfied.

Let x be a point on $\mathcal{S}_1$ and set $q = h/d$, then

$$F(x) = cd^{3-p} \left(\psi_1(q) + \psi_2(q) \right) ,$$

where

$$\psi_1(q) = \int_{y_3 > 1} \left(y_1^2 + y_2^2 + (1 - q - y_3)^2 \right)^{-p/2} dy,$$

$$\psi_2(q) = \int_{y_3 < -1} \left(y_1^2 + y_2^2 + (1 - q - y_3)^2 \right)^{-p/2} dy.$$

Integration is elementary and yields

$$\psi_1(q) = \frac{4\pi}{(p-3)(p-2)} q^{-p+3},$$

$$\psi_2(q) = \frac{4\pi}{(p-3)(p-2)} (2 - q)^{-p+3} .$$

The normal derivative $\partial F / \partial N := \langle F, N \rangle$ of F on $\mathcal{S}_1$ is given by

$$\frac{\partial F}{\partial N} = cd^{2-p} \left(\psi_1'(q) + \psi_2'(q) \right) .$$

We recall that the normal is always directed into the vapor and the constant c is negative. Set

$$f(q) = -cd^{3-p} \left(\psi_1(q) + \psi_2(q) \right) ,$$

then $f(q) > 0$, $f''(q) > 0$ on $0 < q < 2$, and $\lim_{q \to 0} f(q) = \infty$, $\lim_{q \to 2} f(q) = \infty$.

Let κ_0 be the zero of $f'(q) = 0$, i. e., $\kappa_0 = 1$ because of the symmetry of $f(q)$ with respect to $q = 1$. Set

$$X_0 = \exp \left(-\frac{1}{\rho kT} f(1) \right)$$

and assume that the constant X, $0 < X < X_0$, is given, then there are exactly two pairs of planes which are equilibria. These planes are defined by the two zeros κ_1, κ_2 of

$$-\rho kT \ln(X) = f(q) , \quad 0 < \kappa_1 < 1 < \kappa_2 < 2 ,$$

see Fig. 2.2, and the layer with thickness $h = \kappa_1 d$ is stable. We repeat the same considerations on $\mathcal{S}_2$. Let X, $0 < X < X_0$ *be given, then the liquid*

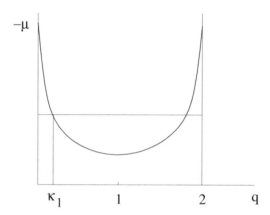

Fig. 2.2   Layer with thickness $h = \kappa_1 d$ is stable

layers with thickness $h = \kappa_1 d$ are stable.

Remark 2.1. The above considerations show that the cavity fills up in a stable manner in contrast to cylinder or ball cavities, where at least one curvature is different from zero, see the following subsections. This contradicts experiments. A modified energy ansatz which takes into account the interaction of molecules in the vapor domain yields probably the right behavior of liquid layers in slit cavities.

From above it follows a result for a single plate: *for any $0 < X < 1$ there is exactly one plane which is an equilibrium, and this equilibrium is stable.* These results for plates and the consideration in [Schiller, Wahab and Mögel (2004)], p. 2230, suggest the following conjecture.

Conjecture 2.1. *If two plates are tilted, i. e., a wedge is occupied by solid material, then there is at least one stable equilibrium which is asymptotically a plane film far away from the edge of the wedge.*

2.2.2   *Adsorption in cylinder cavities*

Let the cross section of the tube be a disk with radius R. The solid domain is given by

$$\Omega_s = \{(y_1, y_2, y_3) \in \mathbb{R}^3 : \ y_1^2 + y_2^2 > R^2, \ -\infty < y_3 < \infty\} \ .$$

We claim that there are stable liquid layers given by $r^2 < x_1^2 + x_2^2 < R^2$, $-\infty < x_3 < \infty$, where $r > 0$, see Fig. 2.3. The potential (2.3) is

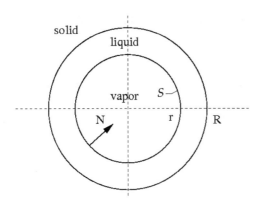

solid

liquid

vapor S

N r R

Fig. 2.3 Cylinder cavity

$$F(x) = cR^{3-p}\psi(q), \ q = r/R \ ,$$

where $r^2 = x_1^2 + x_2^2$, and

$$\psi(q) = \int \left((y_1 - q)^2 + y_2^2 + y_3^2\right)^{-p/2} \, dy \ .$$

The domain of integration is here defined by $y_1^2 + y_2^2 > 1$, $-\infty < y_3 < \infty$.

For stability considerations we need some properties of $\psi(q)$. The function $\psi(q)$ is given by, see [Philip (1977)],

$$\psi(q) = \pi^{3/2} \frac{\Gamma\left(\frac{1}{2}(p-3)\right)}{\Gamma\left(\frac{1}{2}p\right)} \mathbf{F}\left[\frac{1}{2}(p-3), \frac{1}{2}(p-1); 1; q^2\right] \ ,$$

where $\mathbf{F}$ denotes hypergeometric functions. From integral representation of hypergeometric functions, see for instance [Abramowitz and Stegun (1964)], p. 558, Eq. 15.3.1, it follows that $\psi'(q) > 0$ and $\psi''(q) > 0$ on $0 < q < 1$. The definition of $\psi(q)$ implies that $\psi(q) > 0$, $\lim_{q \to 1} \psi(q) = \infty$, $\psi(0) > 0$ and $\psi'(0) = 0$.

The cylinder surface

$$\mathcal{S} = \{(x_1, x_2, x_3) \in \mathbb{R}^3 : \ x_1^2 + x_2^2 = r^2, \ -\infty < x_3 < \infty\}$$

defines an equilibrium if

$$-\mu = \frac{\sigma}{r} - cR^{3-p}\psi\left(\frac{r}{R}\right)$$

is satisfied since the mean curvature of $\mathcal{S}$ is here $H = (2r)^{-1}$. Such a surface is stable if

$$-\frac{\sigma}{r^2} - cR^{2-p}\psi'\left(\frac{r}{R}\right) > 0$$

since the Gauss curvature K of a cylinder surface is zero.

Set

$$f(r) = \frac{\sigma}{r} - cR^{3-p}\psi\left(\frac{r}{R}\right).$$

We recall that the constant c is negative. From the above properties of ψ it follows that $f(r)$ is positive, strictly convex on $0 < r < R$, and $\lim_{r\to 0} f(r) = \infty$, $\lim_{r\to R} f(r) = \infty$, see Fig. 2.4 for a typical graph of $f(r)$.

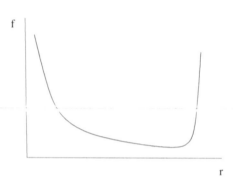

Fig. 2.4   Graph of $f(r)$, layer in a pore

Let r_0 be the zero of $f'(r) = 0$ and set

$$X_0 := \exp\left(-\frac{1}{\rho kT}f(r_0)\right).$$

Then

(i) *for given $0 < X < X_0$ there are exactly two cylinders which are equilibrium states. These cylinders are defined by the zeros r_1, r_2, $r_1 < r_0 < r_2$, of*

$$-\rho kT \ln(X) = f(r).$$

The liquid layer associated to the larger radius r_2 is stable and the other one is not stable. If $X_0 < X \leq 1$, then there is no equilibrium.

(ii) $\lim_{R\to 0} X_0 = 0$.

The behavior (ii) follows from the definition of X_0 and since $f(r)$ tends to infinity, uniformly on $(0, R)$. From the behavior (ii) we see that for given, even small, reduced vapor pressure there are no liquid layers if the pore is sufficiently small.

2.2.3 *Adsorption on a cylinder*

The solid domain is

$$\Omega_s = \{(y_1, y_2, y_3) \in \mathbb{R}^3 : y_1^2 + y_2^2 < R^2, \ -\infty < y_3 < \infty\}$$

and the potential (2.3) is given by

$$F(x) = cR^{3-p}\psi(q), \ q = r/R \ ,$$

where $r^2 = x_1^2 + x_2^2$, $r > R$, and

$$\psi(q) = \int \left((y_1 - q)^2 + y_2^2 + y_3^2\right)^{-p/2} dy \ ,$$

the domain of integration is here $y_1^2 + y_2^2 < 1$, $-\infty < y_3 < \infty$.

The cylinder surface

$$\mathcal{S} = \{(x_1, x_2, x_3) \in \mathbb{R}^3 : x_1^2 + x_2^2 = r^2, \ -\infty < x_3 < \infty\}$$

defines an equilibrium if

$$-\mu = -\frac{\sigma}{r} - cR^{3-p}\psi\left(\frac{r}{R}\right)$$

since the mean curvature of $\mathcal{S}$ is $H = -(2r)^{-1}$. Such a surface is stable if

$$-\frac{\sigma}{r^2} + cR^{2-p}\psi'\left(\frac{r}{R}\right) > 0.$$

From the properties of ψ, we omit the details, it follows that for given $0 < X < 1$ there is exactly one cylinder surface which is an equilibrium, and this is stable, see Fig. 2.5 for a typical graph of $-\sigma/r + |c|R^{3-p}\psi(r/R)$. We recall that the constant c is negative.

2.2.4 *Adsorption in spherical cavities*

The domain occupied by solid material is here

$$\Omega_s = \{(y_1, y_2, y_3) \in \mathbb{R}^3 : y_1^2 + y_2^2 + y_3^2 > R^2\},$$

$R > 0$, and the potential (2.3) is

$$F(x) = cR^{3-p}\psi(q),$$

where

$$\psi(q) := \int \left(y_1^2 + y_2^2 + (y_3 - q)^2\right)^{-p/2} dy.$$

with $q = r/R$ and $r^2 = x_1^2 + x_2^2 + x_3^2$, $0 < r < R$. The domain of integration is $y_1^2 + y_2^2 + y_3^2 > 1$. An elementary calculation yields, if $p > 4$,

$$\psi(q) = \frac{2\pi}{(p-2)(p-3)} \frac{1}{q} \left(\frac{1}{p-4} \left[(1-q)^{-p+4} - (1+q)^{-p+4} \right] \right.$$
$$\left. + \left[(1-q)^{-p+3} - (1+q)^{-p+3} \right] \right).$$

It follows that $\psi'(q) > 0$, $\psi''(q) > 0$, $0 < q < 1$, since $\psi(q)$, $0 < q < 1$, can be written as a convergent power series expansion $\psi(q) = \sum_{k=0}^{\infty} a_{2k} q^{2k}$, where all coefficients a_{2k} are positive.

A sphere $S: x_1^2 + x_2^2 + x_3^2 = r^2$ defines an equilibrium if

$$-\mu = \frac{2\sigma}{r} - cR^{3-p} \psi \left(\frac{r}{R} \right) =: f(r)$$

since the mean curvature of S is $H = r^{-1}$. Such a surface is stable if

$$-\frac{2\sigma}{r^2} - cR^{2-p} \psi' \left(\frac{r}{R} \right) > 0$$

because the Gauss curvature of S is $K = r^{-2}$.

Thus, the same assertions (i) and (ii) of the previous Sec. 2.2 hold for the case of spheres, where f and ψ are taken from above.

Remark 2.2. In contrast to the above result each layer is stable in the volume constrained case. This follows since $S_0: |x| = r$, $0 < r < R$, satisfies the equilibrium condition

$$-2\sigma H - 2\pi |c| R^{3-p} \psi \left(\frac{r}{R} \right) - \lambda = 0,$$

where $H = r^{-1}$, and λ is the constant such that this equation holds for given r. The stability criterion (2.5) is satisfied for all $\xi \in W^{1,2}(S_0) \setminus \{0\}$ which satisfy the side condition $\int_{S_0} \xi \, dA = 0$, since

$$\frac{\partial F}{\partial N} = 2\pi |c| R^{2-p} \psi' \left(\frac{r}{R} \right)$$

is positive and

$$\int_{S_0} \left(|\nabla \xi|^2 - \frac{2}{r^2} \xi^2 \right) dA = \int_{\partial B_1(0)} \left(|\nabla \xi|^2 - 2\xi^2 \right) dA \geq 0$$

for all $\xi \in W^{1,2}(\partial B_1(0))$ which satisfy the side condition $\int_{\partial B_1(0)} \xi \, dA = 0$, where $\partial B_1(0)$ denotes the boundary of the ball $B_1(0)$, see [Wente (1998)]. The proof of this inequality is based on methods from [Wente (1999)].

2.2.4.1 *Adsorption of nitrogen on quartz*

As an example we consider the case of adsorption of nitrogen on quartz for pores of radius $R = a \cdot 10^{-9}$ m, where $2 \le a \le 20$. We assume that $p = 6$ in the potential (2.3), and that

$$T = 77\ K,$$
$$\sigma = 8.88 \cdot 10^{-3}\ Nm^{-1},$$
$$\rho = 16.86 \cdot 10^{27}\ m^{-3},$$
$$k = 1.38065 \cdot 10^{-27}\ NmK^{-1}\ (\text{Boltzmann constant}),$$
$$\mathcal{H} = -10^{-20}\ Nm\ (\text{Hamaker constant}),$$
$$c = \mathcal{H}\pi^{-2} = -0.10132 \cdot 10^{-20}\ Nm.$$

Set $r = x \cdot 10^{-9}$ m, $0 < x < a$, then the above equilibrium condition in this example is given by

$$-1.776 \cdot 10^7 \ln(X) = 10^7 \left(1.776x^{-1} + 0.1013a^{-3}\psi\left(\frac{x}{a}\right)\right).$$

2.2.5 **Adsorption on a sphere**

The domain occupied by solid material is

$$\Omega_s = \{(y_1, y_2, y_3) \in \mathbb{R}^3 : y_1^2 + y_2^2 + y_3^2 < R^2\},$$

$R > 0$, and the potential (2.3)

$$F(x) = cR^{3-p}\psi(q),$$

where

$$\psi(q) = \int \left(y_1^2 + y_2^2 + (y_3 - q)^2\right)^{-p/2} dy$$

with $q = r/R$, and $r^2 = x_1^2 + x_2^2 + x_3^2$, $0 < R < r$. The domain of integration is $y_1^2 + y_2^2 + y_3^2 < 1$. An elementary calculation yields, if $p > 4$,

$$\psi(q) = \frac{2\pi}{(p-2)(p-3)}\frac{1}{q}\left(\frac{1}{p-4}\left[(1+q)^{-p+4} - (q-1)^{-p+4}\right]\right.$$
$$\left. + \left[(q-1)^{-p+3} + (1+q)^{-p+3}\right]\right).$$

A sphere $\mathcal{S}: x_1^2 + x_2^2 + x_3^2 = r^2$, $r > R$, defines an equilibrium if

$$f(r) := -\frac{2\sigma}{r} + |c|R^{3-p}\psi\left(\frac{r}{R}\right) = -\mu$$

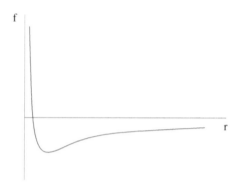

Fig. 2.5   Graph of $f(r)$, adsorption on a sphere

since the mean curvature of $\mathcal{S}$ is $H = -r^{-1}$. Such a surface is stable if

$$-\frac{2\sigma}{r^2} - |c|R^{2-p}\psi'\left(\frac{r}{R}\right) > 0$$

because the Gauss curvature of $\mathcal{S}$ is $K = r^{-2}$.

We obtain qualitatively the same result as in the case of a layer on a cylinder. For given $0 < X < 1$ there is exactly one sphere which is in equilibrium, and this sphere is stable, see Fig. 2.5 for a typical graph of $f(r)$.

2.3   Pores of general geometry

In the case of general geometry of a pore equilibria are not known explicitly. A proposal is to start with an equilibrium surface $\mathcal{S}(\mu_0)$ for a large $|\mu_0|$ calculated by a numerical method, then a numerical continuation procedure yields an equilibrium $\mathcal{S}(\mu)$ for increasing parameter μ. Simultaneously, one checks whether or not $\mathcal{S}(\mu)$ is stable. If $\mathcal{S}(\mu)$ satisfies the stability criterion (2.5) for $\mu < \mu_c$ and $\mathcal{S}(\mu_c)$ violates this condition, then μ_c is said to be the stability bound of this continuation procedure.

Such a numerical continuation was applied to a unilateral problem for the rectangular plate, see [Miersemann and Mittelmann (1990)]. In this example the parameter of continuation is an eigenvalue.

2.4   Stability

Here we consider the question whether or not an equilibrium state defines also strong minimizers of the associated energy functional. In the cases of the considered explicit solutions above each equilibrium defines a strong minimizer of the associated energy provided the second variation is positive for all non vanishing variations. In general, this is not the case as simple counterexamples show, see [Scheeffer (1886)]. The considerations in this section were suggested by the paper of [Wente (2011)] concerning capillary tubes of general cross sections.

2.4.1 Bounded surfaces

Here we suppose that the single cavity $\mathbb{R}^3 \setminus \Omega_s$ is bounded. An example is a spherical cavity.

Theorem 2.3. *Let $\mathbb{R}^3 \setminus \Omega_s$ be a bounded single pore. Suppose that the free interface $\mathcal{S}_0$ of the liquid layer is sufficiently regular, satisfies the necessary condition (2.4), and the second variation is positive for all nonzero variations, see (2.5). Then there exists a $\delta > 0$ such that*

$$\mathcal{E}(\mathcal{S}_0, \mu_0) < \mathcal{E}(\mathcal{S}, \mu_0)$$

for all surfaces $\mathcal{S}$ different from $\mathcal{S}_0$, and located in a δ-neighborhood of $\mathcal{S}_0$. We suppose that $\mathcal{S}$ is sufficiently regular such that the divergence theorem holds.

Proof. The proof is based on a method of [Schwarz (1890)] for minimal surfaces. We will show that an equilibrium surface $\mathcal{S}_0$ can be embedded in a foliation, provided that the second variation is positive for all nonzero variations. Define for a small $\delta > 0$ the δ-neighborhood of $\mathcal{S}_0$ by

$$D_\delta(\mathcal{S}_0) = \{y \in \mathbb{R}^3 : \ y = x(u) + s\, N(u), \ -\delta < s < \delta\}.$$

A family $\mathcal{S}(\mu)$, $\mu \in (\mu_0 - \epsilon, \mu_0 + \epsilon)$, which covers D_δ simply is called *foliation*, and a surface $\mathcal{S}_0$ is called embedded in this family if $\mathcal{S}_0 = \mathcal{S}(\mu_0)$. The existence of such a family is shown in Lemma 2.1 below. Let $x \in D_\delta(\mathcal{S}_0)$ and consider the associated surface $\mathcal{S}(\mu(x))$ from the family $\mathcal{S}(\mu)$. We recall that $\mu(x)$ is constant on $\mathcal{S}(\mu(x))$, and

$$-2\sigma H + F(x) - \mu(x) = 0$$

on $\mathcal{S}(\mu(x))$. Let n be the inward normal on the surface $\mathcal{S}(\mu(x))$ at x, see Fig. 2.6, then

$$\operatorname{div} n = -2H.$$

Combining both equations, we have at $x \in D_\delta(\mathcal{S}_0)$ the equation

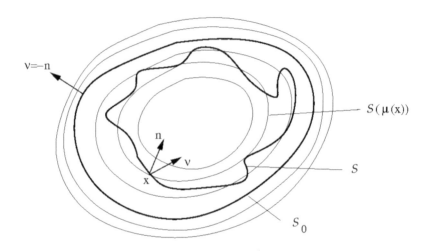

Fig. 2.6　Proof of the theorem

$$\text{div } n + \frac{1}{\sigma}(F(x) - \mu(x)) = 0.$$

Let S be any other regular surface sitting inside of $D_\delta(S_0)$. Here we assume that S is inside of the domain defined by S_0, see Fig. 2.6. For the general case see a remark in [Dierkes, Hildebrandt, Küster and Wohlrab (1992)], p. 81 at the end of the proof of Lemma 1. Let $T_{0,S}$ be the domain between $S_0 \equiv S(\mu_0)$ and S. Then

$$0 = \int_{T_{0,S}} \left(\text{div } n + \frac{1}{\sigma}(F(x) - \mu(x)) \right) dx$$

$$= \int_{\partial T_{0,S}} n \cdot \nu \, dA + \frac{1}{\sigma} \int_{T_{0,S}} (F(x) - \mu(x)) \, dx,$$

where ν is the exterior unit normal on $\partial T_{0,S}$. Since $\nu = -n$ on S_0, it follows

$$|S| - |S_0| = \int_S (1 - n \cdot \nu) \, dA - \frac{1}{\sigma} \int_{T_{0,S}} (F(x) - \mu(x)) \, dx. \qquad (2.7)$$

From the definition (2.1) of the energy we get

$$\mathcal{E}(S, \mu_0) - \mathcal{E}(S_0, \mu_0) = \sigma(|S| - |S_0|) + \int_{T_{0,S}} F(x) \, dx - \mu_0 |T_{0,S}|. \qquad (2.8)$$

Combining formulas (2.8) and (2.7), we find that

$$\mathcal{E}(S, \mu_0) - \mathcal{E}(S_0, \mu_0) = \sigma \int_S (1 - n \cdot \nu) \, dA + \int_{T_{0,S}} (\mu(x) - \mu_0) \, dx.$$

Since $\mu(x) > \mu_0$ if $x \in T_{0,S}$, the theorem is shown. $\qquad\qquad\square$

Lemma 2.1. *Let $\mathcal{S}_0$ be a solution of*

$$-2\sigma H + F - \mu_0 = 0$$

on $\mathcal{S}_0$, and suppose that the second variation of $\mathcal{E}(\mathcal{S}, \mu_0)$ at $\mathcal{S}_0$ is positive on $W^{1,2}(\mathcal{S}_0) \setminus \{0\}$. Then there exists a foliation $\mathcal{S}(\mu)$ of $D_\delta(\mathcal{S}_0)$, where $\delta > 0$ is sufficiently small, and each element of this foliation solves

$$-2\sigma H + F - \mu = 0$$

on $\mathcal{S}(\mu)$.

Proof. We will show that the embedding family is given by $x + \xi(u, \tau)N$, where the scalar function ξ is the solution of

$$-2\sigma H(x + \xi N) + F(x + \xi N) - (\mu_0 + \tau) = 0.$$

The constant τ is from the interval $(-\epsilon, \epsilon)$. Define the mapping

$$M(\xi, \tau): \quad C^{2,\kappa}(\mathcal{S}_0) \times (-\epsilon, \epsilon) \mapsto C^\kappa(\mathcal{S}_0),$$

where $0 < \kappa < 1$ is a constant Hölder exponent, by

$$M(\xi, \tau) = -2\sigma H(x + \xi N) + F(x + \xi N) - (\mu_0 + \tau).$$

We have $M(0, 0) = 0$, and $M_\xi(0, 0)$, defined by

$$M_\xi(0,0)\, h = -\sigma \triangle h - 2\sigma(2H^2 - K)h + \frac{\partial F}{\partial N}h,$$

see [Blaschke (1921)], p. 186, for the formula of the first variation of H, is a regular mapping from $C^{2,\kappa}(\mathcal{S}_0) \mapsto C^\kappa(\mathcal{S}_0)$. It follows from an implicit function theorem, see for example [Kantorowitsch and Akilow (1977)], pp. 660-665, that there is a unique solution

$$\xi = \xi(u, \tau) = \tau v(u) + r(u, \tau)$$

of $M(\xi, \tau) = 0$, where $r(u, \tau) \in C^{2,\kappa}(\mathcal{S}_0)$ $\|r(\tau)\|_{C^{2,\kappa}(\mathcal{S}_0)} = o(\tau)$ as $\tau \to 0$, r_τ exists, $r_\tau \in C^{2,\kappa}(\mathcal{S}_0)$ and $\lim_{\tau \to 0} \|r_\tau(u, \tau)\|_{C^{2,\kappa}(\mathcal{S}_0)} = 0$. Moreover, v is the solution of the Jacobi equation

$$-\sigma \triangle v - 2\sigma(2H^2 - K)v + \frac{\partial F}{\partial N}v - 1 = 0 \qquad (2.9)$$

on $\mathcal{S}_0$. From this equation and from the fact that the second variation is positive for all nonzero variations, we obtain that $v > 0$ on $\mathcal{S}_0$ as follows. Set

$$\zeta(u) = \max\{-v(u), 0\}$$

and

$$q = -2(2H^2 - K) + \frac{1}{\sigma}\frac{\partial F}{\partial N}.$$

Then

$$0 = \int_{S_0} \left(-\triangle v\ \zeta + qv\ \zeta - \frac{1}{\sigma}\zeta \right) dA$$

$$= \int_{S_0} \left(\nabla v \cdot \nabla \zeta + qv\ \zeta - \frac{1}{\sigma}\zeta \right) dA$$

$$= -\int_{\{-v \geq 0\}} \left(|\nabla \zeta|^2 + q\zeta^2 + \frac{1}{\sigma}\zeta \right) dA$$

$$= -\int_{S_0} \left(|\nabla \zeta|^2 + q\zeta^2 + \frac{1}{\sigma}\zeta \right) dA.$$

Combining this equation and inequality (2.5), we find that $\zeta \equiv 0$, i. e., $v \geq 0$ on S_0. This inequality and equation

$$-\triangle v + q\ v = \frac{1}{\sigma}$$

on S_0 imply that $v > 0$ on S_0. If not, then there is a $P \in S_0$ where $v(P) = 0$. Then $-\triangle v = 1/\sigma$ and $\triangle v \geq 0$ at P, which is a contradiction. The previous argument is called *touching principle*. Finally, for given $y = x(u) + s\ N(u) \in D_\delta(S_0)$, $\delta > 0$ sufficiently small, there exists a unique solution $\mu(s)$ of $\xi(u, \mu) = s$ since $\xi_\mu(u, 0) = v(u) > 0$. □

Remark 2.3. Liquid layers in equilibrium on and in spherical cavities studied above which are stable in the sense that the second variation is positive for all nonzero variations are stable in the sense of the previous theorem.

Remark 2.4. In all explicit examples above this lemma is superfluous since an embedding family is defined in a natural way, in the cases of ball or cylinder cavities by varying the radius r. The resulting surfaces are planes, concentric spheres or coaxial cylinders, resp.

2.4.2   Unbounded surfaces

Here we suppose that the single cavity $\mathbb{R}^3 \setminus \Omega_s$ is unbounded. Examples are slit and cylinder cavities. In fact, there are no such cavities. On the other hand, such objects serve as an approximation of long cylinder cavities, for instance. In general, the definition of the energy by (2.1) makes no sense if

$\mathcal{S}$ is unbounded. But the difference (2.8) is well defined if $\mathcal{S}_0$ is replaced by a sufficiently regular bounded sub-domain $\mathcal{M} \subset \mathcal{S}_0$. For given sufficiently small positive δ consider the δ-neighborhood of $\mathcal{M}$

$$D_\delta^{\mathcal{M}}(\mathcal{S}_0) = \{y \in \mathbb{R}^3 : y = x + s\,N, \; x \in \mathcal{M}, \; -\delta < s < \delta\}.$$

Then we study a cylinder type cavity, see Fig. 2.7. We will find an em-

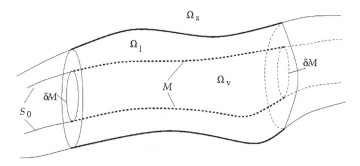

Fig. 2.7   Cylinder type cavity

bedding family $x + \xi(\tau)N$ of $\mathcal{M}$, where $\xi = \xi(u, \tau)$ is the solution of the boundary value problem

$$-2\sigma H(x + \xi N) + F(x + \xi N) - (\mu_0 + \tau) = 0 \quad \text{on } \mathcal{M}$$
$$\xi = \tau \text{ on } \partial\mathcal{M}.$$

Replacing the lateral surfaces Σ of $D_\delta(\mathcal{M})$ by surfaces $\Sigma_0 \subset D_\delta^{\mathcal{M}}(\mathcal{S}_0)$ close to Σ such that $\nu \cdot n = 0$ on Σ_0, where ν is the normal on Σ_0 and n is the normal on the surfaces of the embedding family. The resulting domain is denoted by $D_\delta^0(\mathcal{M})$, see Fig. 2.8. Set $\mathcal{M}_0 = \mathcal{S}_0 \cap D_\delta^0(\mathcal{M})$ and suppose that a sufficiently regular surface $\mathcal{S}$ is in $D_\delta^0(\mathcal{M})$, see Fig. 2.8. Let $T_{0,\mathcal{S}}$ be the domain between $\mathcal{M}_0$ and $\mathcal{S}$. Then the considerations of the proof of the theorem in the previous subsection leads to

Theorem 2.4. *Suppose that the free interface $\mathcal{S}_0$ of the liquid layer is sufficiently regular, satisfies the necessary condition (2.4) and the second variation, see (2.5), is positive for all nonzero variations with compact support. Then there exists a $\delta > 0$ such that*

$$\triangle\mathcal{E}(S, \mathcal{M}_0, \mu_0) := \sigma(|\mathcal{S}| - |\mathcal{M}_0|) + \int_{T_{0,\mathcal{S}}} F(x)\,dx - \mu_0|T_{0,\mathcal{S}}| > 0$$

for all surfaces $\mathcal{S}$ different from $\mathcal{M}_0$, and which are located in $D_\delta^0(\mathcal{M})$.

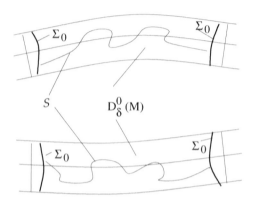

Fig. 2.8   Proof of the theorem

Remark 2.5. In the case of a cylinder cavity, see Sec. 2.2.2, the lateral surfaces are already perpendicular on the surfaces of the embedding family, i. e., we can set $\Sigma = \Sigma_0$.

Remark 2.6. Liquid layers in equilibrium on and in cylindrical cavities studied in Sec. 2.2.2 which are stable in the sense that the second variation is positive for all nonzero variations with compact support are stable in the sense of the previous theorem.

2.5 Appendix

Here we prove formulas which we have used above. Consider a family $\mathcal{S}(\epsilon)$, $|\epsilon| < \epsilon_0$, of surfaces given by

$$z(u, \epsilon) = x(u) + \epsilon \xi(u) N(u),$$

where ξ is a given sufficiently regular function on the parameter domain $U \subset \mathbb{R}^2$ where the support is a *compact* subset of U. Set

$$W(u, \epsilon) = \sqrt{E(u, \epsilon)G(u, \epsilon) - F^2(u, \epsilon)},$$

where $E(u, \epsilon)$, $G(u, \epsilon)$ $F(u, \epsilon)$ are the coefficients of the first fundamental form of $\mathcal{S}(\epsilon)$, and $N(u, \epsilon)$, $H(u, \epsilon)$ are the normal and the mean curvature associated to the surface $\mathcal{S}(\epsilon)$, resp., see Sec. 1.1 concerning the definitions.

Lemma 2.2.

$$\frac{d}{d\epsilon}|\mathcal{S}(\epsilon)| = -2 \int_U H(u, \epsilon)\langle N(u, \epsilon), z_\epsilon(u, \epsilon)\rangle W(u, \epsilon)\, du.$$

Proof. Set $u = (\alpha, \beta)$, then

$$\frac{d}{d\epsilon}|\mathcal{S}(\epsilon)| = \int_U \frac{1}{W(u,\epsilon)} \big(E(u,\epsilon)\langle z_\beta(u,\epsilon), z_{\beta\epsilon}(u,\epsilon)\rangle$$
$$- F(u,\epsilon)[\langle z_\alpha(u,\epsilon), z_{\beta\epsilon}(u,\epsilon)\rangle + \langle z_\beta(u,\epsilon), z_{\alpha\epsilon}(u,\epsilon)\rangle]$$
$$+ G(u,\epsilon)\langle z_\alpha(u,\epsilon), z_{\alpha\epsilon}(u,\epsilon)\rangle \big) \, du.$$

The formula of the lemma follows by integration by parts, see [Dierkes, Hildebrandt, Küster and Wohlrab (1992); Dierkes, Hildebrandt, and Sauvigny (2010)], p. 45, p. 44, resp., and by using the formula $\triangle z = 2HN$, see [Dierkes, Hildebrandt, Küster and Wohlrab (1992); Dierkes, Hildebrandt, and Sauvigny (2010)], p. 71, p. 72, resp., for a proof of this formula. □

Corollary 2.1.

$$\left[\frac{d}{d\epsilon}|\mathcal{S}(\epsilon)|\right]_{\epsilon=0} = -2\int_{\mathcal{S}_0} H_0 \xi \, dA$$

where $\mathcal{S}_0 = \mathcal{S}(0)$*, and* H_0 *denotes the mean curvature of* $\mathcal{S}_0$*.*

Let

$$F(x) = -c \int_{\Omega_s} \frac{dy}{|x - y|^p},$$

where c is a *positive* constant, and $p \geq 4$ is a constant. Set

$$w(\mathcal{S}) = -\int_{\Omega_v(\mathcal{S})} F(x) \, dx,$$

here Ω_v denotes the domain filled with vapor.

Lemma 2.3.

$$\frac{d}{d\epsilon} w(\mathcal{S}(\epsilon)) = \int_U F(z(u,\epsilon))\langle N(u,\epsilon), z_t(u,\epsilon)\rangle W(u,\epsilon) \, du.$$

Proof.

$$\int_{\Omega_v(\mathcal{S}(\epsilon))} F(x) \, dx = \int_{\Omega_v(\mathcal{S}_0)} F(x) \, dx - \int_0^\epsilon \int_U F(z(u,t)) \det \frac{\partial z(u,t)}{\partial(u,t)} \, du \, dt.$$

This formula holds also if $\mathcal{S}(\epsilon)$ is not completely located in the vapor region associated to $\mathcal{S}_0$, see a remark in [Dierkes, Hildebrandt, Küster and Wohlrab (1992)], p. 81. Since

$$\frac{\partial z(u,t)}{\partial(u,t)} = \langle z_\alpha(u,t) \wedge z_\beta(u,t), z_t(u,t)\rangle$$
$$= \langle N(u,t), z_t(u,t)\rangle W(u,t),$$

it follows that

$$\frac{d}{d\epsilon}\int_{\Omega_v(\mathcal{S}(\epsilon))} F(x)\,dx = -\int_U F(z(u,\epsilon))\langle N(u,\epsilon), z_t(u,\epsilon)\rangle W(u,\epsilon)\,du.$$

$\square$

Corollary 2.2.

$$\left[\frac{d}{d\epsilon}w(\mathcal{S}(\epsilon))\right]_{\epsilon=0} = \int_{\mathcal{S}_0} F(x(u))\xi\,dA.$$

Lemma 2.4.

$$\frac{d}{d\epsilon}|\Omega_l(\mathcal{S}(\epsilon))| = \int_U \langle N(u,\epsilon), z_\epsilon(u,\epsilon)\rangle W(u,\epsilon)\,du.$$

Proof. If $\mathcal{S}(\epsilon)$ is completely located in the vapor region associated to $\mathcal{S}_0$, then $\Omega_l(\mathcal{S}(\epsilon)) = \Omega_l(\mathcal{S}_0) \cup \Omega^1$, where $\Omega^1 = \{z(u,t) : \; u \in U, \, 0 < t < \epsilon\}$. The final formula holds for the general case, see a remark in [Dierkes, Hildebrandt, Küster and Wohlrab (1992)], p. 81. Then

$$|\Omega_l(\mathcal{S}(\epsilon))| = |\Omega_l(\mathcal{S}_0)| + \int_0^\epsilon \int_U \langle N(u,t), z_t(u,t)\rangle W(u,t)\,du dt.$$

$\square$

Corollary 2.3.

$$\left[\frac{d}{d\epsilon}|\Omega_l(\mathcal{S}(\epsilon))|\right]_{\epsilon=0} = \int_{\mathcal{S}_0} \xi\,dA.$$

Second derivatives of the integrals with respect to ϵ are used for stability considerations.

Set

$$D(u,\epsilon) = \langle N(u,\epsilon), z_\epsilon(u,\epsilon)\rangle W(u,\epsilon),$$

and let $\triangle$ be the Laplace-Beltrami operator on $\mathcal{S}_0$, and $H_0 = H(u,0)$ and $K_0 = K(u,0)$ the mean and Gauss curvature of $\mathcal{S}_0$, resp.

Lemma 2.5.

$$\left[\frac{d^2}{d\epsilon^2}|\mathcal{S}(\epsilon)|\right]_{\epsilon=0} = -2\int_{\mathcal{S}_0} \xi\left(\triangle\xi + 2(2H_0^2 - K_0)\xi\right)\,dA$$
$$- 2\int_U H_0(u)D_\epsilon(u,0)\,du.$$

Define

$$|\nabla\xi|^2 = \frac{E_0\xi_{u_2}^2 - 2F_0\xi_{u_1}\xi_{u_2} + G_0\xi_{u_1}^2}{W_0^2}.$$

Then the above expansion of $W(\epsilon)$ follows.

(6) Prove the corollary to Lemma 2.2. *Hint.* Partition of unity, and $\mathcal{S}_0$ has no boundary.

Chapter 3

Capillary interfaces

3.1 Governing energy

Let $\Omega_s \subset \mathbb{R}^3$ be a domain occupied by homogeneous solid material. This domain Ω_s is called *container*. By $\Omega_l = \Omega_l(\mathcal{S})$ we denote the domain filled with liquid, where $\mathcal{S}$ denotes the interface between the liquid and the vapor which occupies the domain Ω_v, see Fig. 1.6. Let

$$\mathcal{W}(\mathcal{S}) = \partial\Omega_s \cap \overline{\Omega_l(\mathcal{S})}$$

be the wetted part of the container wall $\partial\Omega_s$. In this chapter we assume that the container wall is a sufficiently smooth surface and that the energy of the problem is given by

$$\mathcal{E}(\mathcal{S}) = \sigma|\mathcal{S}| - \sigma\beta|\mathcal{W}(\mathcal{S})| + \int_{\Omega_l(\mathcal{S})} F(x)\,dx, \qquad (3.1)$$

where F is sufficiently regular. In the following let
$F = Y\rho$,
Y potential energy per unit mass, for example $Y = gx_3$, $g = const.$ gravitational force,
ρ local density of the liquid,
σ surface tension, $\sigma = const. > 0$,
β constant (relative) adhesion coefficient between the fluid and the container wall,
$\mathcal{W}$ wetted part of the container wall,
Ω_l domain occupied by the liquid.

Additionally, we have for given volume V of the liquid the constraint

$$|\Omega_l(\mathcal{S})| = V. \qquad (3.2)$$

By $|\mathcal{S}|$, $|\Omega_l(\mathcal{S})|$ we denote the area resp. the volume of $\mathcal{S}$, $\Omega_l(\mathcal{S})$.

The problem is to find minimizer of $\mathcal{E}$ in an appropriate class of comparison surfaces $\mathcal{S}$.

Let $\mathcal{S}_0$ be a given regular surface. Then we will derive necessary and sufficient conditions such that this surface defines a minimizer of the associated energy functional subject to a given family of comparison surfaces. Assume that $x : u \in U \mapsto \mathbb{R}^3$, $u = (u^1, u^2)$ or $u = (\alpha, \beta)$, defines the regular surface $\mathcal{S}_0$, where $U \subset \mathbb{R}^2$ is a domain such that $x(u) \in \Gamma$ if $u \in \partial U$, and $x(u) \in \mathbb{R}^3 \setminus \{\overline{\Omega_s} \cup \overline{\Omega}\}$ if $u \in U$. We choose the parameters such that the normal

$$N_{\mathcal{S}_0} = \frac{x_\alpha \wedge x_\beta}{|x_\alpha \wedge x_\beta|}$$

is directed out of the liquid, and that the normal N_Γ at the container wall is directed out of the solid material Ω_s of the container, see Fig. 1.6.

3.2 Equilibrium conditions

Following [Finn (1986)], Chap. 1, we define for a given configuration $\mathcal{S}_0$, Σ_0, where $\Sigma_0 = \partial\Omega_0$, a one parameter family of admissible comparison surfaces which are in general not yet volume preserving. Set

$$z(u, \epsilon) = x(u) + \epsilon\zeta(u) + r(u, \epsilon),$$

$0 < \epsilon < \epsilon_0$. We assume that the remainder r is continuously differentiable with respect to all arguments, such that $r = O(\epsilon)$ as $\epsilon \to 0$, $z(u, \epsilon) \in \mathbb{R}^3 \setminus (\overline{\Omega_s} \cup \overline{\Omega_0})$ if $u \in U$, and that $z(u, \epsilon) \in \Gamma$ if $u \in \partial U$ and $z(u, \epsilon) \in \Sigma_0$ if $u \in \partial_2 U$, where

$$\zeta(u) = \xi(u)N_{\mathcal{S}_0}(u) + \eta(u)T_{\mathcal{S}_0}(u)$$

is a given vector field. Here $N_{\mathcal{S}_0}$ denotes the unit normal to $\mathcal{S}_0$ pointed to the exterior of the liquid, and $T_{\mathcal{S}_0}$ is a unit tangent field defined on a strip U_δ of ∂U of width δ, such that on $\partial\mathcal{S}_0$ the vector $\nu_0 := T_{\mathcal{S}_0}$ is orthogonal to $\partial\mathcal{S}_0$ and points to the exterior of the liquid, see Fig. 3.1. In generalization to the normal at the boundary of a two dimensional domain the vector ν_0 at $\partial\mathcal{S}_0$ is called *conormal*. We assume that ξ and η are sufficiently regular on $\overline{U}$, supp $\eta \subset U_\delta$, and $\xi^2 + \eta^2 \leq 1$. Define on $\partial\mathcal{S}_0 = \{x(u) : u \in \partial U\}$ the angle $\gamma \in [0, \pi]$ by $\cos\gamma = N_\Gamma \cdot N_{\mathcal{S}_0}$. This angle depends on $P \in \partial\mathcal{S}_0$. Later on we will see that this angle is a constant if the interface defines an equilibrium. Since $\langle \zeta, N_\Gamma \rangle = 0$ at $\partial\mathcal{S}_0$ and N_Γ, $N_{\mathcal{S}_0}$, $T_{\mathcal{S}_0}$ are in a common plane, it follows that

$$\xi(u) \cos\gamma - \eta(u) \sin\gamma = 0, \tag{3.3}$$

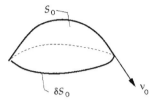

Fig. 3.1   Conormal ν_0 at $\partial \mathcal{S}_0$

where $u \in \partial U$. The family $z(u, \epsilon)$, where ζ satisfies equation (3.3), is called a family of *admissible* comparison surfaces. Such a family exists provided the boundary of the container is sufficiently smooth. The proof exploits an implicit function theorem, we omit the details. Set

$$\mathcal{V} = \{\zeta = \xi N_{\mathcal{S}_0} + \eta T_{\mathcal{S}_0} : (\xi, \eta) \text{ satisfies (3.3)}\}.$$

Suppose that for given $\zeta \in \mathcal{V}$ the family $z(u, \epsilon)$ is volume preserving, i. e., $|\Omega(\mathcal{S}(\epsilon)| = V$ for all $|\epsilon| < \epsilon_0$. It follows that

$$\int_{\mathcal{S}_0} \xi \, dA = 0 \tag{3.4}$$

since, see the corollary to Lemma 3.7 of the appendix to this chapter,

$$\left[\frac{d}{d\epsilon} |\Omega(\mathcal{S}(\epsilon)| \right]_{\epsilon=0} = \int_{\mathcal{S}_0} \xi \, dA.$$

On the other hand, let $\zeta \in \mathcal{V}$ satisfies the side condition (3.4), then there exists a volume preserving family of admissible comparison surfaces given by $z(u, \epsilon) = x(u) + \epsilon\zeta + O(\epsilon^2)$ with a sufficiently regular remainder $O(\epsilon^2)$, see Lemma 3.4 of the appendix to this chapter. Suppose that $\mathcal{S}_0$ is energy minimizing subject to such a family of comparison surfaces $\mathcal{S}(\epsilon)$, i. e., $\mathcal{E}(\mathcal{S}(\epsilon)) \geq \mathcal{E}(\mathcal{S}_0)$, $0 < \epsilon < \epsilon_0$. Then

$$\left[\frac{d}{d\epsilon} \mathcal{E}(\mathcal{S}(\epsilon)) \right]_{\epsilon=0} \geq 0. \tag{3.5}$$

A sufficiently regular surface $\mathcal{S}_0$ defined by $x(u)$ is said to be an *extremal* if inequality (3.5) is satisfied, where $\mathcal{S}(\epsilon)$ is a volume preserving family of comparison surfaces defined above. Set

$$\mathcal{V}_0 = \{\zeta \in \mathcal{V} : \int_{\mathcal{S}_0} \xi \, dA = 0\} \tag{3.6}$$

and

$$\langle f'(x), \zeta \rangle = \left[\frac{d}{d\epsilon} \mathcal{E}(\mathcal{S}(\epsilon)) \right]_{\epsilon=0},$$

$$\langle g'(x), \zeta \rangle = \left[\frac{d}{d\epsilon} |\Omega(\mathcal{S}(\epsilon)| \right]_{\epsilon=0},$$

where

$$\langle g'(x), \zeta \rangle = \int_{S_0} \xi \, dA$$

and, see the corollaries to Lemmas 3.5–3.7 of the appendix to this chapter,

$$\langle f'(x), \zeta \rangle = -2\sigma \int_{S_0} \xi H_0 \, dA + \sigma \int_{\partial S_0} \eta \, ds$$
$$- \sigma \beta \int_{\partial S_0} (\xi \sin \gamma + \eta \cos \gamma) \, ds + \int_{S_0} \xi F(x) \, dA.$$

From a Lagrange multiplier rule, see Chap. 10, it follows that there exists a real constant λ such that

$$\langle f'(x), \zeta \rangle + \lambda \langle g'(x), \zeta \rangle \geq 0 \tag{3.7}$$

for all $\zeta \in \mathcal{V}$.

Theorem 3.1. *Suppose that S is an extremal, then*

$$-2\sigma H + F(x) + \lambda = 0 \quad \text{on } S_0,$$
$$\cos \gamma = \beta \quad \text{on } \partial S_0.$$

Proof. See [Finn (1986)], p. 10. From the above formula (3.7) we get

$$0 = \int_{S_0} (-2\sigma H + F(x) + \lambda) \xi \, dA$$
$$+ \sigma \int_{\partial S_0} [-\beta \xi \sin \gamma + \eta (1 - \beta \cos \gamma)] \, ds$$

for all $(\xi(u), \eta(u))$ satisfying $\xi^2 + \eta^2 \leq 1$ in U and (3.3). Thus

$$-2\sigma H + F(x) + \lambda = 0 \quad \text{on } S_0$$

and

$$\int_{\partial S_0} [-\beta \xi \sin \gamma + \eta (1 - \beta \cos \gamma)] \, ds \geq 0$$

for all $(\xi(\alpha, \beta), \eta(\alpha, \beta))$, $(\alpha, \beta) \in \partial U$, satisfying (3.3). Set

$$\xi = \tau \sin \gamma, \quad \eta = \tau \cos \gamma,$$

where $\tau = \tau(\alpha, \beta)$, $|\tau| \leq 1$, $\tau \in C(\partial U)$. Then

$$\int_{\partial S_0} \tau (\beta - \cos \gamma) \, ds \leq 0$$

for all these τ. It follows that $\cos \gamma = \beta$ on ∂S_0. $\qquad \square$

transforms this problem into

$$\operatorname{div} Tv = Bv \qquad \text{in } \Omega,$$
$$\nu \cdot Tv = \cos \gamma \qquad \text{on } \partial\Omega,$$

where

$$B = \kappa a^2$$

is called *Bond number*. Here we assume that the mapping $y = x/a$ maps $\Omega(a)$ onto a fixed domain Ω, independent of a.

3.3 Explicit solutions

There are only few examples where explicit solutions are known. We will see in Chap. 7 that such solutions can serve as a leading term for asymptotic expansions of solutions to problems where no explicit solutions are known.

3.3.1 *Ascent of a liquid at a vertical wall*

Let $\Omega = \{(x_1, x_2) \in \mathbb{R}^2 : x_1 > 0\}$ be the right half plane, and consider the boundary value problem

$$\operatorname{div} Tu = \kappa u \qquad \text{in } \Omega,$$
$$\nu \cdot Tu = \cos \gamma \qquad \text{on } \partial\Omega,$$

where κ is a positive constant. Now we assume *additionally* that there exists a solution u which depends on $x = x_1$ only such that $\lim_{x \to \infty} u(x) = 0$ and $\lim_{x \to \infty} u'(x) = 0$. In the following we will find such a solution. From the maximum principle of [Finn and Hwang (1989)], see Sec. 7.1, it follows that this solution is the only one of the original boundary value problem *without the additionally assumptions* at the infinity. Thus we have to consider a boundary value problem for an ordinary differential equation.

$$\left(\frac{u'(x)}{\sqrt{1 + u'(x)^2}} \right)' = \kappa u(x), \qquad 0 < x < \infty, \qquad (3.11)$$

$$\lim_{x \to 0,\, x > 0} \frac{u'(x)}{\sqrt{1 + u'(x)^2}} = -\cos\gamma, \qquad (3.12)$$

$$\lim_{x \to \infty} u(x) = 0, \qquad \lim_{x \to \infty} u'(x) = 0. \qquad (3.13)$$

From the identities

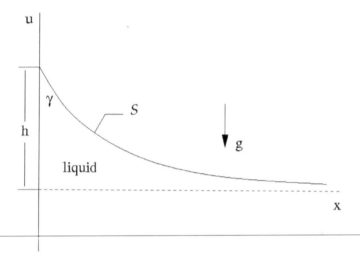

Fig. 3.3   Ascent of liquid at a vertical wall

$$\left(\frac{u'(x)}{\sqrt{1+u'(x)^2}}\right)' = \frac{u''(x)}{(1+u'(x)^2)^{3/2}} \tag{3.14}$$

$$= -\frac{1}{u'(x)}\left(\frac{1}{\sqrt{1+u'(x)^2}}\right)', \tag{3.15}$$

here we assume at this step that $u'(x) \neq 0$, it follows from (3.11) that

$$-\frac{1}{u'(x)}\left(\frac{1}{\sqrt{1+u'(x)^2}}\right)' = \kappa u.$$

Then

$$\left(\frac{1}{\sqrt{1+u'(x)^2}}\right)' = -\frac{1}{2}\kappa(u^2)'. \tag{3.16}$$

for all $x \in (0,d)$. Consequently, we have

$$\frac{1}{\sqrt{1+u'(x)^2}} = -\frac{1}{2}\kappa u^2 + C,$$

where C is a constant. From the boundary conditions (3.13) we see that $C = 1$. Then

$$\frac{1}{\sqrt{1+u'(x)^2}} = 1 - \frac{1}{2}\kappa u^2. \tag{3.17}$$

Combining this formula with the boundary condition (3.12), we get for the ascent $h = u(0)$ of the liquid at the wall, see Fig. 3.3 in the case $0 \leq \gamma \leq \pi/2$, that

$$h = \sqrt{\frac{2}{\kappa}}\sqrt{1 - \sin\gamma}.$$

Instead of $u = u(x)$ we consider the inverse function $x = x(u)$. Here we assume that $x'(u) \neq 0$. It turns out that this inequality is satisfied for the solution which we will calculate in the following. Then we obtain from (3.17) the equation

$$-\frac{x'(u)}{\sqrt{1 + x'(u)^2}} = 1 - \frac{1}{2}\kappa u^2. \tag{3.18}$$

We have used the additional assumption that $x'(u) < 0$, which is satisfied for the calculated solution. We suppose here that $0 \leq \gamma < \pi/2$. From (3.18) it follows

$$x'(u) = \frac{\kappa u^2/2 - 1}{\sqrt{1 - (\kappa u^2/2 - 1)^2}}$$

$$= \frac{u^2/\alpha - 1}{u/\alpha\sqrt{2\alpha - u^2}},$$

where $\alpha := 2/\kappa$. Using the substitution

$$\tau = \sqrt{2\alpha - u^2},$$

we find that

$$x(u) = \int \frac{\kappa u^2/2 - 1}{\sqrt{1 - (\kappa u^2/2 - 1)^2}}\, du + C$$

$$= -\tau + \alpha \int \frac{d\tau}{2\alpha - \tau^2} + C$$

$$= -\sqrt{2\alpha - u^2} + \frac{\alpha}{2\sqrt{2\alpha}}\ln\left(\frac{\sqrt{2\alpha} + \sqrt{2\alpha - u^2}}{\sqrt{2\alpha} - \sqrt{2\alpha - u^2}}\right) + C.$$

The constant follows from the condition $x(u(0)) = 0$, where $u(0) = h$, and the final formula is given by

$$x(u) = \sqrt{2\alpha - h^2} - \sqrt{2\alpha - u^2} + \sqrt{\frac{\alpha}{2}}\ln\left(\frac{(\sqrt{2\alpha} - \sqrt{2\alpha - h^2})u}{(\sqrt{2\alpha} - \sqrt{2\alpha - u^2})h}\right),$$

where $\alpha = 2/\kappa$.

Remark 3.3. Caused by the strong non-linearity of the equation the information about the behavior of the solution at infinity is contained in the equation itself.

3.3.2 *Ascent of a liquid between two parallel plates*

Let $\Omega(d) = \{(x_1, x_2) \in \mathbb{R}^2 : 0 < x_1 < d\}$, $d > 0$, and consider the boundary value problem

$$\mathrm{div}\ Tu = \kappa u \qquad \text{in } \Omega,$$
$$\nu \cdot Tu = \cos \gamma_1 \qquad \text{on } \Gamma_1,$$
$$\nu \cdot Tu = \cos \gamma_2 \qquad \text{on } \Gamma_2,$$

where κ is a positive constant, and Γ_1, Γ_2 are the two walls, see Fig. 3.4. Now we assume *additionally* that there exists a solution u which depends on $x = x_1$ only, see Fig. 3.4. As in the previous example, from the maximum

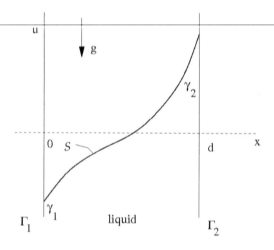

Fig. 3.4   Ascent of liquid between two vertical walls

principle of [Finn and Hwang (1989)], see Sec. 7.1, it follows that this solution is the only one of the original boundary value problem *without the additionally assumption*. Thus we have to consider boundary value problem

$$\left(\frac{u'(x)}{\sqrt{1 + u'(x)^2}} \right)' = \kappa u(x), \qquad 0 < x < d, \tag{3.19}$$

$$\lim_{x \to 0+0} \frac{u'(x)}{\sqrt{1 + u'(x)^2}} = -\cos \gamma_1, \tag{3.20}$$

$$\lim_{x \to d-0} \frac{u'(x)}{\sqrt{1 + u'(x)^2}} = \cos \gamma_2. \tag{3.21}$$

As in the case of the ascent at a vertical wall we get from the formula (3.16) the equation

$$\sin \gamma_2 - \sin \gamma_1 = -\frac{1}{2}\kappa \left(u^2(d) - u^2(0) \right).$$

Remark 3.4. Concerning the numerical solution of the boundary value problem (3.19)–(3.21) see Chap. 7.

3.4 Zero gravity solutions

We consider the zero gravity problem

$$\operatorname{div} Tu = 2H \quad \text{in } \Omega, \tag{3.22}$$

$$\nu \cdot Tu = \cos \gamma \quad \text{on } \partial\Omega, \tag{3.23}$$

where H is a constant. There are domains where a solution exists and also domains without a solution.

3.4.1 *Examples*

Here we consider some explicit examples.

3.4.1.1 *Tube, circular cross section*

Let $\Omega(a) = B_a(0)$ be a disk with radius a and center at the origin. We assume that the solution $u(x)$ in consideration is rotationally symmetric with center at the origin, i. e., $u(x) = v(r)$, where $r = \sqrt{x_1^2 + x_2^2}$, then

$$\frac{1}{r}\left(\frac{rv'(r)}{\sqrt{1+v'(r)^2}} \right)' = 2H, \tag{3.24}$$

$$\lim_{r \to a-0} \frac{v'(r)}{\sqrt{1+v'(r)^2}} = \cos \gamma. \tag{3.25}$$

From the maximum principle of [Concus and Finn (1974)], see Sec. 7.1, it follows that every solution must be rotationally symmetric. Integrating (3.22) over $\Omega(a)$, and using the boundary condition, we get

$$2H = \frac{2\cos \gamma}{a}. \tag{3.26}$$

Equation (3.24) implies that

$$\frac{rv'(r)}{\sqrt{1+v'(r)^2}} = Hr^2 + C.$$

Setting $r = 0$, it follows that the constant C is zero. Then, if $0 \leq \gamma \leq \pi/2$,

$$v(r) = -\sqrt{\frac{1}{H^2} - r^2} + C$$

with another constant C. Thus $v(r)$ defines a circle with radius $1/|H|$ and center at $(0, C)$. Suppose that the liquid of given volume V occupies a domain as shown in Fig. 3.5a, i. e., we consider a tube closed by a flat bottom, and suppose that there is enough liquid such there is no dry spot on the bottom. Then we find from

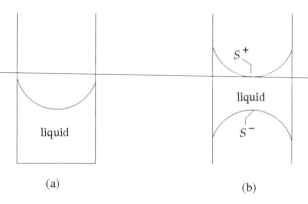

(a)   (b)

Fig. 3.5   Liquid in a circular tube

$$V = 2\pi \int_0^a r v(r) \, dr$$

that

$$C = \frac{V + 2\pi a^2/3}{\pi a^2}.$$

Assume that there is no bottom, and let $\mathcal{S}^+$ and $\mathcal{S}^-$ be defined by

$$v^+ = -\sqrt{\frac{1}{H^2} - r^2} + C^+, \quad v^- = \sqrt{\frac{1}{H^2} - r^2} + C^-,$$

resp., with constants C^+, C^-, see Fig. 3.5b. Then

$$C^+ - C^- = \frac{V + 4\pi a^2/3}{\pi a^2}.$$

Here we assume that there is enough liquid such that $\mathcal{S}^+$ is above of $\mathcal{S}^-$.

3.4.1.2 *Liquid between two coaxial cylinders*

Consider problem (3.22) and (3.23) over an annular domain $0 < a < |x| < b < \infty$ and assume that there exists a solution which is defined as a graph over the x-plane, $x = (x_1, x_2)$, then

$$\text{div } Tu = 2H \qquad \text{in } a < |x| < b,$$
$$\nu \cdot Tu = \cos \gamma_1 \qquad \text{on } |x| = a,$$
$$\nu \cdot Tu = \cos \gamma_2 \qquad \text{on } |x| = b$$

Scaling $x = by$, $v(y) = u(by)/b$ leads to

$$\text{div } Tv = 2A \qquad \text{in } q < |y| < 1,$$
$$\nu \cdot Tv = \cos \gamma_1 \qquad \text{on } |y| = q,$$
$$\nu \cdot Tv = \cos \gamma_2 \qquad \text{on } |y| = 1,$$

where $q = a/b$. Integrating the differential equation, we obtain for the constant

$$A = \frac{\cos \gamma_2 + q \cos \gamma_1}{1 - q^2}.$$

Suppose that the solution in consideration is rotationally symmetric. From a maximum principle, see Sec. 7.1, it follows that every solution must be rotationally symmetric. Set $w(r) = v(y)$, $r = \sqrt{y_1^2 + y_2^2}$, see Fig. 3.6. Then

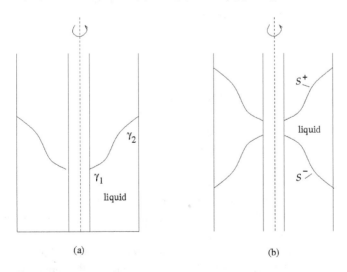

(a) (b)

Fig. 3.6 Liquid between two coaxial cylinders

the boundary value problem is given by

$$\frac{1}{r}\left(\frac{rw'(r)}{\sqrt{1+w'(r)^2}}\right)' = 2A, \quad q < r < 1,$$

$$\lim_{r\to q+0}\frac{w'(r)}{\sqrt{1+w'(r)^2}} = -\cos\gamma_1,$$

$$\lim_{r\to 1-0}\frac{w'(r)}{\sqrt{1+w'(r)^2}} = \cos\gamma_2.$$

Elementary calculations lead to the elliptic integral

$$w(r) = \int_q^r \frac{f(s)}{\sqrt{1-f^2(s)}}\, ds + const.,$$

where

$$f(s) = \frac{1}{s}\left(-q\cos\gamma_1 + A(s^2 - q)\right),$$

see an exercise. In Fig. 3.6b the upper surface $\mathcal{S}^+$ is given by $bw(|y|) + B$, and the lower surface $\mathcal{S}^-$ by $-bw(|y|) + C$ with a constant C.

3.4.1.3   *Cross section is a regular n-gon*

Consider the boundary value problem (3.22) and (3.23), where $0 \le \gamma \le \pi/2$. Let Ω be a regular n-gon, where the corners lie on the unit circle. It is easily seen that the lower hemisphere given by

$$v_0(x;\gamma) = -\frac{1}{H}\sqrt{1-H^2|x|^2},$$

with

$$H = \frac{\cos\gamma}{\cos(\pi/n)},$$

is a solution of the boundary value problem (3.22) and (3.23) if $H \le 1$, see Fig. 3.7 for the case that Ω is a square. The previous inequality implies that $\pi/n \le \gamma$, or, equivalently, $\alpha + \gamma \ge \pi/2$, where 2α denotes the interior angle at the corners. From a maximum principle, see Sec. 7.1, it follows that every solution of (3.22) and (3.23) is given by $v_0 + const.$, provided that $\pi/n < \gamma$ holds. The following theorem shows that there is no solution of (3.22) and (3.23) if $\pi/n > \gamma$.

Let $\Omega \subset \mathbb{R}^2$ be a domain with a corner located at the origin and assume that this corner with opening angle $0 < 2\alpha < \pi$ is locally defined by straight lines, see Fig. 3.8.

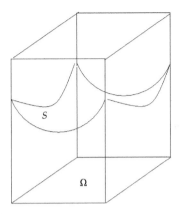

Fig. 3.7   Cross section is a square

Theorem 3.2 (Concus and Finn, 1970). *Suppose that* $u \in C^2(\Omega_a) \cap C^1(\overline{\Omega_a} \setminus \{0\})$ *is a solution of*

$$div\, Tu = f(x, u) \qquad in\ \Omega_a,$$
$$\nu \cdot Tu = \cos \gamma \qquad on\ \Sigma_a^{(1)} \cup \Sigma_a^{(2)},$$

where $f(x, u)$ is a given function satisfying

$$\sup_{x \in \Omega_a} |f(x, u(x))| < \infty.$$

Then

$$\alpha + \gamma \geq \frac{\pi}{2}.$$

Proof. Integrating the differential equation over Ω_c, $0 < c \leq a$, we find that

$$\int_{\Sigma_c^{(1)} \cup \Sigma_c^{(2)}} \nu \cdot Tu\, ds + \int_{\Gamma_c} \nu \cdot Tu\, ds = \int_{\Omega_c} f(x, u(x))\, dx,$$
$$2|\Sigma_c| \cos \gamma + \int_{\Gamma_c} \nu \cdot Tu\, ds = O(|\Omega_c|).$$

Since $|\nu \cdot Tu| \leq 1$, it follows

$$2|\Sigma_c| \cos \gamma \leq |\Gamma_c| + O(c^2),$$
$$2c \frac{\cos \gamma}{\cos \alpha} \leq 2c \tan \alpha + O(c^2),$$

which implies that $\cos \gamma \leq \sin \alpha$, and the theorem is shown. We recall that $0 \leq \gamma \leq \pi/2$. □

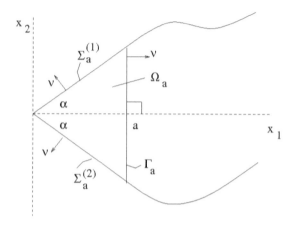

Fig. 3.8   Corner domain

Corollary 3.1. *If there is a corner of Ω with interior angle 2α and defined locally through straight lines. If $\alpha + \gamma < \pi/2$ holds, then there is no bounded solution of*

$$\operatorname{div} Tu = f(x, u) \qquad in\ \Omega,$$
$$\nu \cdot Tu = \cos\gamma \qquad on\ \partial\Omega,$$

where $f(x, u)$ is given function satisfying

$$\sup_{x \in \Omega, |u| \le C} |f(x, u)| = m(C) < \infty.$$

Thus one expects that liquid, oil for example, flows out of a glass with edges of interior angles less than $\pi - 2\gamma$. On the other hand an upper edge can serve as a wetting barrier which prevents the liquid to flow out of the glass, see Chap. 6.

3.4.2   *A necessary condition for the existence of a solution*

The following necessary condition was shown in [Concus and Finn (1974)], see also [Finn (1986)], p. 135. Consider a subdivision of the bounded domain Ω through a curve Γ, see Fig. 3.9, and set

$$H_\gamma = \frac{\Sigma}{\Omega} \cos\gamma.$$

Then we have

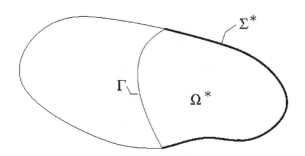

Fig. 3.9   Necessary condition

Theorem 3.3 (Concus and Finn, 1974). *A necessary condition for the existence of a solution of the boundary value problem (3.22) and (3.23) is that the functional*

$$\Phi[\Omega^*] := \Gamma - \Sigma^* \cos\gamma + H_\gamma \Omega^*$$

is positive for every $\Omega^* \neq \emptyset$, Ω.

Proof. Integrating equation (3.22) over Ω and using the boundary condition (3.23), we get

$$\text{div } Tu = \frac{\Sigma}{\Omega} \cos\gamma.$$

Integrating this equation over Ω^* and using the boundary condition on Σ^*, we find that

$$\left(\frac{\Sigma}{\Omega}\Omega^* - \Sigma^*\right)\cos\gamma = \int_\Gamma \nu \cdot Tu \, ds.$$

The assertion of the theorem follows since $|Tu| \leq 1$. $\qquad\square$

Example 3.1. Assume that the cross section has a narrow slit as drawn in Fig. 3.10. If the slit is sufficiently small then there is no solution of the associated boundary value problem provided the contact angle is below of a fixed angle. Consider a circle with center at the corner and with a radius δ larger then the width ϵ of the slit, see Fig 3.10. An elementary calculation shows that

$$\Gamma = 2\pi\delta\frac{\alpha}{2\pi} + 2\pi\delta\frac{\beta}{2\pi},$$

$$\Sigma^* = 2\delta + 2\sqrt{\delta^2 - \epsilon^2},$$

$$= 2\delta + 2\delta\sqrt{1 - \left(\frac{\epsilon}{\delta}\right)^2},$$

$$\Omega^* = O(\delta^2)$$

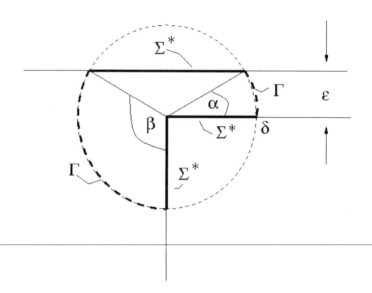

Fig. 3.10   Narrow slit, notations

since $\beta = \alpha + \pi/2$ and $\sin \alpha = \epsilon/\delta$.

Let $\epsilon = \delta^2$, then $\alpha = \delta + O(\delta^3)$ and

$$\Sigma^* = 4\delta + O(\delta^3),$$
$$\Gamma = \frac{\pi}{2}\delta + O(\delta^2).$$

Consequently,

$$\Phi[\Omega^*] = \gamma - \Sigma^* \cos \gamma + H_\gamma \Omega^*$$
$$= \left(\frac{\pi}{2} - 4\cos\gamma\right)\delta + O(\delta^2).$$

Thus $\Phi[\Omega^*]$ is negative if the contact angle γ satisfies

$$\gamma < \gamma_0 := \arccos\left(\frac{\pi}{8}\right),$$

provided that δ, $\delta > 0$, is sufficiently small.

3.4.3   *A sufficient condition for the existence of a solution*

The following useful result was shown by [Chen (1980)].

Theorem 3.4 (Chen, 1980). *Suppose that Ω satisfies the interior disk condition with radius Ω/Σ, see Fig. 3.11. Then there exists a solution for every contact angle γ, $0 \le \gamma \le \pi$.*

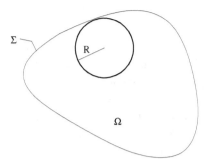

Fig. 3.11   Sufficient condition

Remark 3.5. Since the contact angle changes from time to time caused by pollution or vibration of the container, the above result is of some practical interest.

Example 3.2. Consider a cross slit as drawn in Fig. 3.12. Since

$$\frac{\Omega}{\Gamma} = \frac{2(h_1 + h_2 + h_3 + h_4) + 4 + 2\pi}{2(h_1 + h_2 + h_3 + h_4) + 4\pi}$$

is less than 1. Thus there exist a solution for every contact angle.

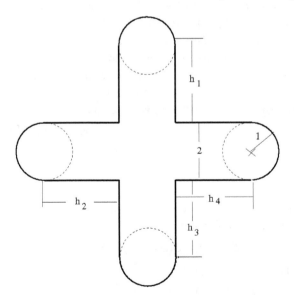

Fig. 3.12   Cross slit

3.5 Stability

We will show that a capillary interface in equilibrium defines a strong energy minimizer if there exists an embedding foliation of capillary surfaces and if the Lagrange multiplier decreases with increasing volume of liquid in the container. We also give a proof of Wente's second variation formula which includes the borderline cases that the contact angle is 0 or π. Finally, we discuss the question of the existence of an embedding foliation for a given capillary interface in equilibrium.

Let K_Γ be the curvature at $P \in \partial S_0$ of the plane curve defined by the intersection of Γ with the plane through P and the linear span N_{S_0} (at P) and ν_0. This curvature is considered as non-negative if the curve bends in the direction of N_Γ. Extending an idea of Finn, we form a class of volume preserving admissible comparison surfaces $S(\epsilon)$ defined by

$$z(u, \epsilon) = x(u) + \epsilon\zeta(u) + q(\epsilon)\xi_0(u)N_0(u) + \frac{\epsilon^2}{2}|\zeta(u)|^2 r(u) + O(\epsilon^3), \quad (3.27)$$

where $\zeta \in V_0$ is given, and

$$r(u) = K_\Gamma(u)N_\Gamma(u)\rho(u).$$

Here $K_\Gamma(u)$ and $N_\Gamma(u)$ denote smooth extensions of K_Γ and N_Γ, which are defined on ∂S_0, to U_δ, and $\rho(u)$ is a given smooth function which satisfies $\rho(u) = 1$ if $u \in \partial U$ and $\rho(u) = 0$ if $\text{dist}(u, \partial U) > \delta$, δ small. For example, $N_\Gamma(u) := N_{S_0}(u)\cos\gamma - T_{S_0}(u)\sin\gamma$ is an extension of $N_\Gamma(u)$, $u \in \partial U$ to U_δ. We can assume that the remainder $O(\epsilon^3)$ is sufficiently regular, which follows from an implicit function theorem, we omit the details. The function ξ_0 is fixed with $\text{supp}\,\xi_0 \subset U$ and $\int_{S_0} \xi_0 dA = 1$, and the function $q(\epsilon)$ satisfies $q(0) = 0$ and $q'(0) = 0$, see Lemma 3.4. We recall that

$$V = \{\zeta = \xi N_{S_0} + \eta T_{S_0} : (\xi, \eta) \text{ satisfies } (3.3)\}$$

and

$$V_0 = \{\zeta \in V : \int_{S_0} \xi\, dA = 0\}.$$

Suppose that $S_0 = S(0)$ is an equilibrium interface, i. e., $x(u)$ satisfies the necessary conditions of Theorem 3.1. Set

$$\mathcal{L}(S, \lambda) = \mathcal{E}(S) + \lambda(|\Omega(S)| - V),$$

then, if $\zeta \in V_0$,

$$\mathcal{E}(S(\epsilon)) - \mathcal{E}(S_0) = \mathcal{L}(S(\epsilon), \lambda) - \mathcal{L}(S_0, \lambda)$$

$$= \epsilon \left[\frac{d}{d\epsilon}\mathcal{L}(S(\epsilon), \lambda)\right]_{\epsilon=0} + \frac{\epsilon^2}{2}\left[\frac{d^2}{d\epsilon^2}\mathcal{L}(S(\epsilon), \lambda)\right]_{\epsilon=0} + O(\epsilon^3).$$

Since

$$\left[\frac{d}{d\epsilon}\mathcal{L}(\mathcal{S}(\epsilon),\lambda)\right]_{\epsilon=0} = 0$$

for all $\zeta \in \mathcal{V}$, see the Lagrange equation (3.7), in particular for $\zeta \in \mathcal{V}_0$, then

$$\left[\frac{d^2}{d\epsilon^2}\mathcal{L}(\mathcal{S}(\epsilon),\lambda)\right]_{\epsilon=0} \geq 0$$

is a necessary condition such that the inequality

$$\mathcal{E}(\mathcal{S}(\epsilon)) \geq \mathcal{E}(\mathcal{S}_0)$$

holds for all $|\epsilon| < \epsilon_0$.

Next we will calculate

$$\left[\frac{d^2}{d\epsilon^2}\mathcal{L}(\mathcal{S}(\epsilon),\lambda)\right]_{\epsilon=0}$$

under the assumption that $\mathcal{S}_0$ defines an equilibrium, i. e., it satisfies the equations of Theorem 3.1.

From Lemma 3.5 and Lemma 3.7 of the appendix to this chapter we see that

$$\begin{aligned}
\frac{d}{d\epsilon}\mathcal{L}(\mathcal{S}(\epsilon),\lambda) = &-2\sigma \int_U H(u,\epsilon)\langle N(u,\epsilon), z_\epsilon(u,\epsilon)\rangle W(u,\epsilon)\, du \\
&+ \sigma \int_{\partial\mathcal{S}(\epsilon)} \langle \nu(u,\epsilon), z_\epsilon(u,\epsilon)\rangle\, ds(\epsilon) \\
&- \sigma\beta \int_{\partial U} \langle Z_{0,\epsilon}(\tau,\epsilon), \overline{\nu}(u,\epsilon)\rangle |Z_{0,\tau}(\tau,\epsilon)|\, d\tau \\
&+ \int_U F(z(u,\epsilon)\langle N(u,\epsilon), z_\epsilon(u,\epsilon)\rangle W(u,\epsilon)\, du \\
&+ \lambda \int_U \langle N(u,\epsilon), z_\epsilon(u,\epsilon)\rangle W(u,\epsilon)\, du,
\end{aligned} \tag{3.28}$$

where $ds(\epsilon) = |Z_\tau(\tau,\epsilon)|d\tau$, $Z(\tau,\epsilon) = z(u(\tau),\epsilon)$, $Z_0(\tau,t) = z_0(u(\tau),t)$ and $u(\tau)$ is a regular parameter representation of ∂U.

Set

$$D(u,\epsilon) = \langle N(u,\epsilon), z_\epsilon(u,\epsilon)\rangle W(u,\epsilon),$$

then

$$\left[\frac{d^2}{d\epsilon^2}\mathcal{L}(\mathcal{S}(\epsilon),\lambda)\right]_{\epsilon=0} = -2\sigma \int_U H_\epsilon(u,0)D(u,0)\, du$$

$$- 2\sigma \int_U H(u,0)D_\epsilon(u,0)\, du$$

$$+ \sigma \int_{\partial U} \left[\frac{\partial}{\partial\epsilon}\langle\nu(u,\epsilon), z_\epsilon(u,\epsilon)\rangle\right]_{\epsilon=0} |Z_\tau(\tau,0)|\, d\tau$$

$$+ \sigma \int_{\partial U} \langle\nu(u,0), z_\epsilon(u,0)\rangle \left[\frac{\partial}{\partial\epsilon}|Z_\tau(\tau,\epsilon)|\right]_{\epsilon=0} d\tau$$

$$- \sigma\beta \int_{\partial U} \left[\frac{\partial}{\partial\epsilon}\langle\overline{\nu}(u,\epsilon), Z_{0,\epsilon}(\tau,\epsilon)\rangle\right]_{\epsilon=0} |Z_{0,\tau}(\tau,\epsilon)|\, d\tau$$

$$- \sigma\beta \int_{\partial U} \langle\overline{\nu}(u,0), Z_{0,\epsilon}(\tau,0)\rangle \left[\frac{\partial}{\partial\epsilon}|Z_{0,\tau}(\tau,\epsilon)|\right]_{\epsilon=0} d\tau$$

$$+ \int_U \langle\nabla F(x(u)), z_\epsilon(u,0)\rangle D(u,0)\, du$$

$$+ \int_U F(x(u))D_\epsilon(u,0)\, du$$

$$+ \lambda \int_U D_\epsilon(u,0)\, du.$$

From the differential equation $-2\sigma H_0(u) + F(x(u)) + \lambda = 0$ on $\mathcal{S}_0$ and the formula, see Lemma 3.11 of the appendix to this chapter,

$$2H_\epsilon(u,0) = \triangle\xi(u) + 2(2H_0^2(u) - K_0(u))\xi(u) + 2H_{0,\alpha}(u)\eta^\alpha(u),$$

where H_0, K_0 are the mean and Gauss curvature of $\mathcal{S}_0$, resp., ξ and η^α are defined through $\zeta = \xi N_0 + \eta^\alpha x_\alpha$, $\eta^\alpha x_\alpha = \eta\nu_0$ on $\mathcal{S}_0$, and ξ, η satisfy $\xi\cos\gamma - \eta\sin\gamma$ on ∂U, we get

$$\left[\frac{d^2}{d\epsilon^2}\mathcal{L}(\mathcal{S}(\epsilon),\lambda)\right]_{\epsilon=0} = -\sigma \int_{\mathcal{S}_0} (\triangle\xi + 2(2H_0^2 - K_0)\xi)\,\xi\, dA$$

$$+ \int_{\mathcal{S}_0} \langle\nabla F, N_0\rangle\xi^2\, dA$$

$$+ \sigma \int_{\partial U} \left[\frac{\partial}{\partial\epsilon}(\langle\nu(u,\epsilon), z_\epsilon(u,\epsilon)\rangle - \beta\langle\overline{\nu}(u,\epsilon), Z_{0,\epsilon}(\tau,\epsilon)\rangle)\right]_{\epsilon=0}$$

$$|Z_{0,\tau}(\tau,0)|\, d\tau$$

$$+ \sigma \int_{\partial U} (\langle\nu(u,0), z_\epsilon(u,0)\rangle - \beta\langle\overline{\nu}(u,0), Z_{0,\epsilon}(\tau,0)\rangle)$$

$$\left[\frac{\partial}{\partial\epsilon}|Z_{0,\tau}(\tau,\epsilon)|\right]_{\epsilon=0} d\tau\ .$$

Here we have used that $D(u,0) = \xi W(u,0)$, $Z_\tau(\tau,0) = Z_{0,\tau}(\tau,0)$,

$$z_\epsilon(u,0) = z_{0,\epsilon}(u,0) = \zeta = \xi N_0 + \eta^\alpha x_\alpha,$$

and

$$-2\sigma H_{0,\alpha}(u) + \langle \nabla F(x(u)), x_\alpha(u) \rangle = 0$$

on $\mathcal{S}_0$.

Next we show that the integrands of the last term in the previous formula for the second derivative of $\mathcal{L}$ vanish. We have

$$\langle \nu(u,0), z_\epsilon(u,0) \rangle - \beta \langle \overline{\nu}(u,0), Z_{0,\epsilon}(\tau,0) \rangle = \langle \nu_0 - \beta\overline{\nu}, \zeta \rangle$$

since $z_\epsilon(u,0) = \zeta$ and $Z_{0,\epsilon}(\tau,0) = \zeta$ on $\partial\mathcal{S}$. At $\partial\mathcal{S}_0$ the vectors ν_0, $\overline{\nu}_0$, N_Γ are in a common plane, and $\overline{\nu}_0$ and N_Γ, N_{Σ_0}, resp., are orthogonal. Thus we have at $\partial\mathcal{S}_0$, with scalar functions a, b, the formula

$$\nu_0 = a\overline{\nu}_0 + bN_\Gamma.$$

Using the boundary condition $\cos\gamma = \beta$ on $\partial\mathcal{S}_0$ it follows

$$\nu_0 - \overline{\nu}_0 \cos\gamma = -N_\Gamma \sin\gamma. \tag{3.29}$$

Then the brackets vanish since $\zeta \perp N_\Gamma$ at $\partial\mathcal{S}_0$. This is a consequence of formula

$$\langle \zeta, \nu_0 - \beta\overline{\nu}_0 \rangle = \tau(\cos\gamma - \beta),$$

where τ is a continuous function defined on ∂U, see the proof of Theorem 3.1.

For the brackets in the integrands of the remaining boundary integrals we get from formula (3.29) above that

$$\left[\frac{\partial}{\partial\epsilon}(\langle \nu(u,\epsilon), z_\epsilon(u,\epsilon) \rangle - \beta\langle \overline{\nu}(u,\epsilon), Z_{0,\epsilon}(\tau,\epsilon) \rangle) \right]_{\epsilon=0}$$
$$= \langle \nu_0, z_{\epsilon\epsilon}(u,0) \rangle - \beta\langle \overline{\nu}_0, Z_{0,\epsilon\epsilon}(\tau,0) \rangle + \langle \nu_\epsilon, z_\epsilon(u,0) \rangle - \beta\langle \overline{\nu}_\epsilon(u,0), \zeta \rangle$$
$$= \langle K_\Gamma|\zeta|^2 N_\Gamma, \nu_0 - \beta\overline{\nu}_0 \rangle + \langle \zeta, \nu_\epsilon(u,0) - \beta\overline{\nu}_\epsilon(u,0) \rangle$$
$$= -K_\Gamma|\zeta|^2 \sin\gamma + \langle \zeta, \nu_\epsilon(u,0) - \beta\overline{\nu}_\epsilon(u,0) \rangle.$$

Finally we obtain

$$\langle \zeta, \nu_\epsilon(u,0) - \beta\overline{\nu}_\epsilon(u,0) \rangle = \langle \xi N_0 + \eta\nu_0, \nu_\epsilon(u,0) \rangle - \beta\langle \zeta, \overline{\nu}_\epsilon(u,0) \rangle$$
$$= \xi\langle N_0, \nu_\epsilon(u,0) \rangle$$
$$= \xi\left(\frac{\partial\xi}{\partial\nu_0} + K_{\mathcal{S}_0}\eta \right).$$

Here we have used that $\langle \bar{\nu}_\epsilon(u,0), \bar{\nu}(u,0) \rangle = 0$, $\zeta = a\bar{\nu}_0$ with a scalar function a, and the formula, see Lemma 3.9 of the appendix to this chapter,

$$\langle N_0, \nu_\epsilon(u,0) \rangle = \frac{\partial \xi}{\partial \nu_0} + K_{\mathcal{S}_0} \eta,$$

where $K_{\mathcal{S}_0}$ is the curvature at $P \in \partial \mathcal{S}_0$ of the plane curve defined through the intersection of $\mathcal{S}_0$ and the plane through P which is spanned by $N_{\mathcal{S}_0}$ at P and $\nu_0 := \nu(0)$ at P. The curvature $K_{\mathcal{S}_0}$ is considered as non-negative if the curve in consideration bends in the direction of $N_{\mathcal{S}_0}$ at P.

Summarizing, we obtain after integration by parts, see [Dierkes, Hildebrandt, Küster and Wohlrab (1992); Dierkes, Hildebrandt, and Sauvigny (2010)], p. 45, p. 44, resp.,

Lemma 3.1. *Suppose that $x(u)$ which defines $\mathcal{S}_0$ satisfies the equations of Theorem 3.1. Then*

$$\left[\frac{d^2}{d\epsilon^2} \mathcal{L}(\mathcal{S}(\epsilon), \lambda) \right]_{\epsilon=0} = \sigma \int_{\mathcal{S}_0} \left(|\nabla \xi|^2 - 2(2H_0^2 - K_0)\xi^2 \right) \, dA$$

$$+ \int_{\mathcal{S}_0} \langle \nabla F, N_0 \rangle \xi^2 \, dA$$

$$+ \sigma \int_{\partial_1 \mathcal{S}_0} \left(\xi \eta K_{\mathcal{S}_0} - |\zeta|^2 K_\Gamma \sin \gamma \right) \, ds \, .$$

Corollary 3.2. *Suppose that γ_k are different from 0 or π, then*

$$\left[\frac{d^2}{d\epsilon^2} \mathcal{L}(\mathcal{S}(\epsilon), \lambda) \right]_{\epsilon=0} = \sigma \int_{\mathcal{S}_0} \left(|\nabla \xi|^2 - 2(2H_0^2 - K_0)\xi^2 \right) \, dA$$

$$+ \int_{\mathcal{S}_0} \langle \nabla F, N_0 \rangle \xi^2 \, dA$$

$$+ \sigma \int_{\partial \mathcal{S}_0} \left(\frac{\cos \gamma}{\sin \gamma} K_{\mathcal{S}_0} - \frac{1}{\sin \gamma} K_\Gamma \right) \xi^2 \, ds.$$

Proof. The formula follows since

$$|\zeta|^2 = \langle \xi N_0 + \eta \nu_0, \xi N_0 + \eta \nu_0 \rangle$$
$$= \xi^2 + \eta^2$$

on $\partial \mathcal{S}$, and

$$\xi \cos \gamma - \eta \sin \gamma = 0$$

on $\partial \mathcal{S}_0$. $\qquad\qquad\qquad\qquad\qquad\qquad\qquad\qquad\qquad\qquad\qquad \square$

Remark 3.6. We recall that $2H_0^2 - K_0 \geq 0$ holds for every regular surface.

Remark 3.7. The second variation formula of Lemma 3.1 was derived by [Wente (1966)] under the assumption that γ is different from 0 or π. In the case of constant mean curvature, i. e., if $F = 0$, a proof was given in [Ros and Souam (1997)], also for contact angles different from 0 or π. Our proof includes the borderline cases $\gamma = 0$ and $\gamma = \pi$ as well as the case of a non-vanishing F. At some points and notations our proof is close to that of [Ros and Souam (1997)], but we do not assume the existence of a smooth continuation of the given surface $\mathcal{S}_0$ across its boundary.

An extremal is often called *stable* by definition if the second variation, given by the formula of the above lemma, is positive for all non-vanishing ξ satisfying

$$\int_{\mathcal{S}_0} \xi \, dA = 0.$$

3.5.1 *Strong minimizers*

Let $\mathcal{S}_0$ be a sufficiently regular bounded capillary interface in equilibrium, i. e., $\mathcal{S}_0$ satisfies the equilibrium conditions of Theorem 3.1. Define the Lagrange functional $\mathcal{L}(\mathcal{S}, \lambda)$ for sufficiently regular admissible interfaces $\mathcal{S}$ by

$$\mathcal{L}(\mathcal{S}, \lambda) = \mathcal{E}(\mathcal{S}) + \lambda(|\Omega_l(\mathcal{S})| - V_0),$$

where $\mathcal{E}$ is given by (3.1), $\lambda \in \mathbb{R}$ and $V_0 = |\Omega_l(\mathcal{S}_0)|$.

Admissible means that $\partial \mathcal{S} \subset \Gamma$ and $\mathcal{S} \subset \mathbb{R}^3 \setminus \overline{\Omega_s}$. If $\mathcal{S}$ is volume preserving, i. e., $|\Omega_l(\mathcal{S})| = V_0$, then

$$\mathcal{L}(\mathcal{S}, \lambda) = \mathcal{E}(\mathcal{S}).$$

Define for a small $\delta > 0$ a δ-neighborhood D_δ of $\overline{\mathcal{S}_0}$ by

$$D_\delta = \{x \in \mathbb{R}^3 \setminus \Omega_s : \text{ dist}(x, \overline{\mathcal{S}_0}) < \delta\}.$$

Let $\mathcal{S}(\tau)$, $|\tau| < \epsilon$, be a family of sufficiently regular and admissible surfaces given by $z(u, \tau) \in \mathbb{R}^3$, $u \in U$, where $U \subset \mathbb{R}^2$ is a fixed parameter domain. A family $\mathcal{S}(\tau)$ which covers D_δ simply is called a *foliation*, and the surface $\mathcal{S}_0$ is called *embedded* in this family if $\mathcal{S}_0$ is defined by $z(u, 0)$.

Remark 3.8. The method of foliation was used by [Wente (2011)] to prove a result concerning capillary tubes of non-constant circular cross sections.

Assumption 3.1. (i) There exists an embedding foliation of S_0 defined by

$$2\sigma H = \lambda(\tau) + F(z(u, \tau)) \quad \text{in } S(\tau),$$
$$N_{S(\tau)} \cdot N_\Gamma = \cos \gamma \quad \text{on } \partial S(\tau),$$

where H is the mean curvature of $S(\tau)$ at $z(u, \tau)$.
(ii) $|\Omega(S(\tau))|$ is increasing and $\lambda(\tau)$ is decreasing with growing τ.

Theorem 3.5. *Suppose that S_0 satisfies the previous assumptions (i) and (ii). Then $\mathcal{E}(S) \geq \mathcal{E}(S_0)$ for all admissible volume preserving comparison surfaces $S \subset D_\delta$.*

Proof. Let $x \in D_\delta$ and consider the associated surface $S(\tau(x))$ from the family $S(\tau)$. We recall that $\tau(x)$ is constant on $S(\tau(x))$ and

$$2\sigma H = \lambda(\tau(x)) + F \quad \text{in } S(\tau(x)), \tag{3.30}$$
$$N_{S(\tau(x))} \cdot N_\Gamma = \cos \gamma \quad \text{on } \partial S(\tau(x)). \tag{3.31}$$

Let n be the normal on the surface $S(\tau(x))$ at x. We assume that this normal is directed out of the liquid, see Fig. 3.13. Then

$$\operatorname{div} n = -2H \tag{3.32}$$

at x, where H is the mean curvature of $S(\tau(x))$ at x. For a proof of formula (3.32) see [Dierkes, Hildebrandt, Küster and Wohlrab (1992)], p. 77, for instance. Combining (3.30) and (3.32), we get

$$\operatorname{div} n + \frac{1}{\sigma}(\lambda(\tau(x)) + F) = 0$$

at $x \in D_\delta$. Let T^+ and T^- be the domains enclosed by S_0 and S which are above or below, resp., of S_0. See Fig. 3.13 for further notations used in the following. Integrating over T^+ and T^-, resp., and using the divergence theorem

$$\int_{T^+} \operatorname{div} n \, dx = \int_{\partial T^+} n \cdot \nu \, dA, \tag{3.33}$$

we obtain

$$0 = -|\Sigma^+| \cos \gamma - |S_0^+| + \int_{S^+} n \cdot \nu \, dA + \frac{1}{\sigma} \int_{T^+} (F + \lambda(\tau(x))) \, dx$$

$$= -|\Sigma^+| \cos \gamma - |S_0^+| + |S^+| + \int_{S^+} (n \cdot \nu - 1) \, dA$$

$$+ \frac{1}{\sigma} \int_{T^+} (F + \lambda(\tau(x))) \, dx$$

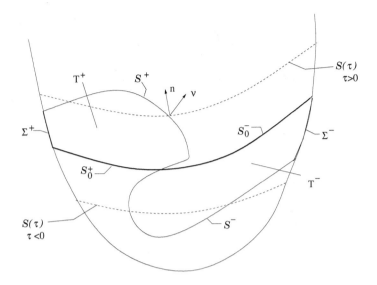

Fig. 3.13   Proof of Theorem 3.3

and

$$0 = -|\Sigma^-|\cos\gamma + |\mathcal{S}_0^-| - |\mathcal{S}^-| + \int_{\mathcal{S}^-} (n\cdot\nu + 1)\, dA + \frac{1}{\sigma}\int_{T^-} (F + \lambda(\tau(x)))\, dx$$

Consequently, we have

$$|\mathcal{S}| - |\mathcal{S}_0| - (|\Sigma^+| - |\Sigma^-|)\cos\gamma = \int_{\mathcal{S}^+} (1 - n\cdot\nu)\, dA + \int_{\mathcal{S}^-} (n\cdot\nu + 1)\, dA$$

$$+ \frac{1}{\sigma}\int_{T^-} (F + \lambda(\tau(x)))\, dx - \frac{1}{\sigma}\int_{T^+} (F + \lambda(\tau(x)))\, dx.$$

Since

$$|\mathcal{W}(\mathcal{S})| - |\mathcal{W}(\mathcal{S}_0)| = |\Sigma^+| - |\Sigma^-|$$

and

$$\mathcal{L}(\mathcal{S}, \lambda_0) - \mathcal{L}(\mathcal{S}_0, \lambda_0) = \sigma\left(|\mathcal{S}| - |\mathcal{S}_0| - \beta(|\mathcal{W}(\mathcal{S})| - |\mathcal{W}(\mathcal{S}_0)|)\right)$$

$$+ \int_{\Omega_l(\mathcal{S})} F\, dx - \int_{\Omega_l(\mathcal{S}_0)} F\, dx + \lambda_0\left(|\Omega_l(\mathcal{S})| - |\Omega_l(\mathcal{S}_0)|\right),$$

it follows that

$$\mathcal{L}(\mathcal{S}, \lambda_0) - \mathcal{L}(\mathcal{S}_0, \lambda_0) = \sigma\left(\int_{\mathcal{S}^+} (1 - n\cdot\nu)\, dA + \int_{\mathcal{S}^-} (n\cdot\nu + 1)\, dA\right)$$

$$+ \int_{T^-} (F + \lambda(\tau(x)))\, dx - \int_{T^+} (F + \lambda(\tau(x)))\, dx$$

$$+ \int_{\Omega_l(\mathcal{S})} F\, dx - \int_{\Omega_l(\mathcal{S}_0)} F\, dx + \lambda_0\left(|\Omega_l(\mathcal{S})| - |\Omega_l(\mathcal{S}_0)|\right).$$

Thus

$$\mathcal{L}(\mathcal{S}, \lambda_0) - \mathcal{L}(\mathcal{S}_0, \lambda_0) = \sigma \left(\int_{\mathcal{S}^+} (1 - n \cdot \nu) \, dA + \int_{\mathcal{S}^-} (n \cdot \nu + 1) \, dA \right)$$
$$+ \int_{T^-} (\lambda(\tau(x)) - \lambda_0) \, dx + \int_{T^+} (\lambda_0 - \lambda(\tau(x))) \, dx. \qquad (3.34)$$

Since $\mathcal{S}$ is volume preserving by assumption we have

$$\mathcal{L}(\mathcal{S}, \lambda_0) - \mathcal{L}(\mathcal{S}_0, \lambda_0) = \mathcal{E}(\mathcal{S}) - \mathcal{E}(\mathcal{S}_0).$$

We get from (3.34) that $\mathcal{E}(\mathcal{S}) \geq \mathcal{E}(\mathcal{S}_0)$. $\qquad\qquad\qquad\qquad\qquad\square$

Example 3.3. As an example we consider the capillary tube with bounded cross section $\Omega \subset \mathbb{R}^2$. Assume that the capillary interface $\mathcal{S}_0$ is given by a graph $z = u(x)$ over the x-plane, $x = (x_1, x_2)$. Then

$$\operatorname{div} Tu = \kappa \, u + \lambda_0 \qquad \text{in } \Omega,$$
$$\nu \cdot Tu = \cos \gamma \qquad \text{on } \partial\Omega,$$

where $Tu = \nabla u / \sqrt{1 + |\nabla u|^2}$, ν is the exterior unit normal at $\partial\Omega$ and $\kappa > 0$ is the capillarity constant. Let V_0 be the given volume. A foliation is explicitly given by

$$\mathcal{S}(\tau): \ z(x, \tau) = u(x) + \tau,$$

where $\tau \in \mathbb{R}$, $|\tau| < \epsilon$. Then $V(\tau) = V_0 + \tau|\Omega|$ is the volume of liquid under $\mathcal{S}(\tau)$, and the Lagrange parameter is

$$\lambda_0(\tau) = \frac{1}{|\Omega|} \left(|\partial\Omega| \cos \gamma - \kappa V(\tau) \right).$$

Both assumptions (i), (ii) are satisfied.

3.5.2   *On the existence of an embedding foliation*

Let $\mathcal{S}_0$ be a sufficiently regular surface defined by $x = x(u)$ satisfying the equilibrium equations of Theorem 3.1. Let τ be a real parameter in a neighborhood of 0. Let $\mathcal{S}(\tau)$ be an admissible family of surfaces, defined by

$$z(u, \tau) = x(u) + \zeta(u, \tau),$$

where

$$\zeta(u, \tau) = \xi(u, \tau)N_{\mathcal{S}_0} + \eta(u, \tau)T_{\mathcal{S}_0},$$

with supp $\eta(u, \tau) \subset U_\delta \times (-\tau_0, \tau_0)$ for a $\tau_0 > 0$, and U_δ is a closed boundary strip of U as introduced in Sec. 3.3. Admissible means, see Sec. 3.1, that

$z(u, \tau)$ is in the interior of the container if $u \in U$ and on the container wall if $u \in \partial U$.

Assume that $S(\tau)$ is sufficiently regular and satisfies

$$2\sigma H = \lambda(\tau) + F(z(u, \tau)) \quad \text{in } S(\tau), \tag{3.35}$$

$$N_{S(\tau)} \cdot N_\Gamma = \cos\gamma \quad \text{on } \partial S(\tau), \tag{3.36}$$

$$|\Omega_l(S(\tau))| = V_0 + \tau, \tag{3.37}$$

where H denotes the mean curvature of $S(\tau)$ at $z(u, \tau)$, N_Γ is the normal on the container wall Γ directed as shown in Fig. 1.6, and $V_0 = |\Omega_l(S_0))|$ is the given volume.

Suppose that

$$\xi(u, \tau) = \xi_1(u)\tau + \frac{1}{2}\xi_2(u)\tau^2 + O(\tau^3),$$

$$\eta(u, \tau) = \eta_1(u)\tau + \frac{1}{2}\eta_2(u)\tau^2 + O(\tau^3),$$

$$\lambda(\tau) = \lambda_0 + \lambda_1\tau + O(\tau^2),$$

where the coefficients and the remainders are sufficiently regular.

Set

$$\zeta_l(u) = \xi_l(u)N_{S_0} + \eta_l(u)T_{S_0}, \quad l = 1, 2,$$

then

$$z(u, \tau) = x(u) + \zeta_1(u)\tau + \frac{1}{2}\zeta_2(u)\tau^2 + O(\tau^3).$$

By assumption, $z(u, \tau)$ is on the container wall if $u \in \partial U$. Consequently, we have at ∂U that

$$\xi_1(u)\cos\gamma - \eta_1(u)\sin\gamma = 0, \tag{3.38}$$

$$\zeta_2(u) = K_\Gamma|\zeta_1|^2 N_\Gamma. \tag{3.39}$$

Then we get from equation (3.35) and Lemma 3.11 of the appendix (first variation of mean curvature) that on S_0

$$\sigma(\triangle\xi_1 + 2(2H^2 - K)\xi_1 + 2H_{,\alpha}\eta w^\alpha) = \lambda_1 + \frac{\partial F}{\partial N}\xi_1 + \langle \nabla_x F, T \rangle \eta_1,$$

is satisfied, where $T = w^\alpha x_{,\alpha}$, and $H = H(u)$ is the mean curvature of S_0 at $x(u)$. Using the first order necessary condition

$$2\sigma H = \lambda_0 + F(x(u)) \quad \text{on } S_0,$$

the above equation reduces to

$$\sigma(\triangle\xi_1 + 2(2H^2 - K)\xi_1) = \lambda_1 + \frac{\partial F}{\partial N}\xi_1.$$

From (3.38) we obtain that $\xi_1 = 0$ on ∂U if $\gamma = 0$ or $\gamma = \pi$. If $0 < \gamma < \pi$, then we find from the above boundary condition (3.36), equations (3.38), (3.39) and the corollary to Lemma 3.11 of the appendix to this chapter that

$$\frac{\partial \xi_1}{\partial \nu} + \left(\frac{\cos \gamma}{\sin \gamma} K_{\mathcal{S}} - \frac{1}{\sin \gamma} K_\Gamma \right) \xi_1 = 0$$

on ∂S_0. Finally equation (3.37) implies that

$$\int_{S_0} \xi_1 \, dA = 1.$$

Set

$$p = -2(2H^2 - K) + \frac{1}{\sigma} \frac{\partial F}{\partial N},$$

$$q = \frac{\cos \gamma}{\sin \gamma} K_{\mathcal{S}} - \frac{1}{\sin \gamma} K_\Gamma.$$

~~Consequently, $(\xi, \lambda) := (\xi_1, \lambda_1)$ is a solution of~~

$$-\triangle \xi + p\xi + \frac{\lambda}{\sigma} = 0 \qquad \text{in } S_0, \tag{3.40}$$

$$\frac{\partial \xi}{\partial \nu} + q\xi = 0 \quad \text{on } \partial S_0 \text{ if } 0 < \gamma < \pi, \tag{3.41}$$

$$\xi = 0 \quad \text{on } \partial S_0 \text{ if } \gamma = 0, \text{ or } \gamma = \pi, \tag{3.42}$$

$$\int_{S_0} \xi \, dA = 1, \tag{3.43}$$

provided there exists a sufficiently regular family of admissible surfaces.

On the other hand, there exists a unique solution (ξ, λ) of the system (3.40)–(3.43) if the second variation is positive. Set

$$Q(\phi, \psi) = \int_{S_0} (\nabla \phi \cdot \nabla \psi + p\, \phi\psi) \, dA + \int_{\partial S_0} q\, \phi\psi \, ds \ \text{ if } 0 < \gamma < \pi,$$

$$Q(\phi, \psi) = \int_{S_0} (\nabla \phi \cdot \nabla \psi + p\, \phi\psi) \, dA \ \text{ if } \gamma = 0, \text{ or } \gamma = \pi.$$

Lemma 3.2. *Suppose that $0 < \gamma < \pi/2$. Then there exists a unique solution (ξ, λ) of (3.40)–(3.43), provided the second variation $Q(\phi, \phi)$ is positive for all $\phi \in W^{1,2}(S_0) \setminus \{0\}$ satisfying $\int_{S_0} \phi \, dA = 0$ if $0 < \gamma < \pi/2$. If $\gamma = 0$ or $\gamma = \pi$ we have to replace $W^{1,2}$ through $W_0^{1,2}$.*

Proof. For the convenience of the reader we will give a proof. Set $J(\phi) = Q(\phi, \phi)/2$ and consider the minimum problem $\min \ J(\phi)$, where the minimum is taken over all $\phi \in W^{1,2}(S_0)$ or $\phi \in W_0^{1,2}(S_0)$, resp., satisfying the side condition $\int_{S_0} \phi \, dA = 1$. Take a sufficiently regular ϕ_0 which satisfies the previous side condition and set $\phi = \psi + \phi_0$. Then the above minimum problem changes to min $J(\psi + \phi_0)$, where the minimum is taken over all $\psi \in W^{1,2}(\mathcal{S}_0)$ or $\psi \in W_0^{1,2}(\mathcal{S}_0)$, resp., satisfying the side condition

$$\int_{\mathcal{S}_0} \psi \, dA = 0.$$

The existence of a solution ψ_0 follows from the assumption concerning the second variation. Consequently, there exists a Lagrange multiplier λ such that $(\psi_0 + \phi_0)$ is a solution of (3.40)–(3.43). The uniqueness is also a consequence of the positivity of the second variation. Assume (ξ_1, λ_1) and (ξ_2, λ_2) are solutions of (3.40)–(3.43), then $\xi := \xi_1 - \xi_2$ satisfies (3.40)–(3.42), with $\lambda = \lambda_1 - \lambda_2$, and the homogeneous side condition $\int_{\mathcal{S}_0} \xi \, dA = 0$. It follows that $Q(\xi, \xi) = 0$. Thus $\xi = 0$, which implies $\lambda_1 = \lambda_2$, see the above differential equation (3.40) with $\lambda = \lambda_1 - \lambda_2$. $\qquad\square$

Lemma 3.3. *Assume that $\lambda_1 < 0$. Let $0 < \gamma < \pi$ and if the second variation $Q(\phi, \phi)$ is positive for all $\phi \in W^{1,2}(\mathcal{S}_0) \backslash \{0\}$ satisfying $\int_{\mathcal{S}_0} \phi \, dA = 0$, then $\xi_1(u) > 0$ on $\overline{U}$.*

If $\gamma = 0$ or $\gamma = \pi$ and the second variation $Q(\phi, \phi)$ is positive for all $\phi \in W_0^{1,2}(\mathcal{S}_0) \backslash \{0\}$ satisfying $\int_{\mathcal{S}_0} \phi \, dA = 0$, then $\xi_1(u) > 0$ on (the open set) U.

Proof. Set $\xi = \xi_1$ and $\lambda = \lambda_1/\sigma$ in this proof. Let $\xi = \xi^+ + \xi^-$, where $\xi^+ = \max\{\xi, 0\}$, $\xi^- = \min\{\xi, 0\}$, and assume that $\xi < 0$ on a subset of U of positive measure. Then

$$\int_{\mathcal{S}_0} \xi^- \, dA < 0.$$

Define $\xi_0 = \xi^+ + \alpha\xi^-$, where

$$\alpha = 1 - \frac{1}{\int_{\mathcal{S}_0} \xi^- \, dA}.$$

Then $\alpha > 1$ and $\int_{\mathcal{S}_0} \xi_0 \, dA = 0$. We have, see (3.40) and (3.41),

$$Q(\xi, \phi) + \lambda \int_{\mathcal{S}_0} \Phi \, dA = 0 \tag{3.44}$$

for all $\phi \in W^{1,2}(\mathcal{S}_0)$. Inserting $\phi = \xi_0$ und $\xi = \xi_0 + (1 - \alpha)\xi^-$, we see that

$$0 = Q(\xi_0, \xi_0) + (1 - \alpha)Q(\xi^-, \xi_0)$$
$$= Q(\xi_0, \xi_0) + (1 - \alpha)\alpha Q(\xi^-, \xi^-)$$

since $\xi_0 = \xi^+ + \alpha\xi^-$. Hence $Q(\xi^-, \xi^-) > 0$ because of $\xi_0 \neq 0$ and $\alpha > 1$. On the other hand, if we insert $\phi = \xi^-$ in equation (3.44), we get

$$Q(\xi^-, \xi^-) + \lambda \int_{S_0} \xi^- \, dA = 0,$$

which is a contradiction if $\lambda \leq 0$. Thus we have shown $\xi(u) \geq 0$ in U, provided that $\lambda \leq 0$. If $\lambda < 0$, then it follows from the differential equation (3.40) and the strong maximum principle that $\xi(u) > 0$ in U. Next we show that $\xi(u) > 0$ on $\overline{U}$, provided that $0 < \gamma < \pi$. Let $\xi(P) = 0$ at $P \in \partial U$. Then we will show that

$$\frac{\partial \xi}{\partial \nu} > 0 \tag{3.45}$$

at P, which is a contradiction to the boundary condition (3.41). In fact, the Hopf boundary point lemma, [Hopf (1952)], says that $\partial\xi/\partial n > 0$, where $n = (n_1, n_2)$ is the exterior normal at ∂U. We recall that ∂U is sufficiently smooth by assumption. Since $\partial\xi/\partial n = \xi_{,\alpha}n^\alpha$ and $\partial\xi/\partial\nu = \xi_{,\alpha}\nu^\alpha$, where $\nu^\alpha = \sqrt{g}g^{\alpha\beta}n_\beta$, see [Dierkes, Hildebrandt, Küster and Wohlrab (1992)], p. 44. Then

$$\nu^\alpha n_\alpha = \sqrt{g}g^{\alpha\beta}n_\alpha n_\beta > 0$$

since $(g^{\alpha\beta})$ is positive definite. This inequality says that $\nu = (\nu^1, \nu^2)$ considered as a vector at P in $\mathbb{R}^2$ makes an angle with n which is less than π. Thus at P we have $\nu = a\,n + b\,t$, where $a > 0$. Here t denotes a unit tangent vector at ∂U at P. Then $\partial\xi/\partial\nu = a\,\partial\xi/\partial n + b\,\partial\xi/\partial t = a\,\partial\xi/\partial n$ at P, which proves inequality (3.45). $\square$

The above two lemmas suggest the following conjecture.

Conjecture 3.1. *Suppose that $\lambda_1 < 0$, and that the second variation is positive for variations which are not identically zero and satisfy the side condition $\int_{S_0} \phi \, dA = 0$. Then there exists an embedding foliation of admissible surfaces.*

3.6   Appendix

Here we prove formulas which are used in the previous sections of this chapter. We consider a family of admissible comparison surfaces given by

$$z_0(u, \epsilon) = x(u) + \epsilon\zeta(u) + r(u, \epsilon),$$

$|\epsilon| < \epsilon_0$. We assume that the remainder r is continuously differentiable with respect to all arguments, such that $r = O(\epsilon)$ as $\epsilon \to 0$. Moreover let

$z_0(u, \epsilon) \in \mathbb{R}^3 \setminus (\overline{\Omega_s} \cup \overline{\Omega_0})$ if $u \in U$, $z_0(u, \epsilon) \in \Gamma$ if $u \in \partial U$. The vector field $\zeta(u)$ is defined by

$$\zeta(u) = \xi(u)N_{\mathcal{S}_0}(u) + \eta(u)T_{\mathcal{S}_0}(u).$$

Here $N_{\mathcal{S}_0}$ denotes the unit normal to $\mathcal{S}_0$ pointed to the exterior of the liquid, and $T_{\mathcal{S}_0}$ is a unit tangent field defined on closed strip U_δ of ∂U of width δ, such that on $\partial \mathcal{S}_0$, $\nu_0 := T_{\mathcal{S}_0}$ is orthogonal to $\partial \mathcal{S}_0$ and points to the exterior of the liquid, see Fig. 3.1. We assume that ξ and η are sufficiently regular on $\overline{U}$, supp $\eta \subset U_\delta$, $\xi^2 + \eta^2 \leq 1$, and $\langle \zeta, N_\Gamma \rangle = 0$, i. e., ξ, η satisfy (3.3). By $\nu(u, \epsilon)$, resp. $\overline{\nu}(u, \epsilon)$, we denote the exterior normal to $\partial \mathcal{S}(\epsilon)$ in $\mathcal{S}(\epsilon)$,

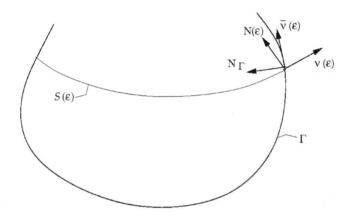

Fig. 3.14   Normals in consideration

resp., in the container wall Γ, see Fig. 3.14. At $P \in \partial \mathcal{S}(\epsilon)$ the vectors N_Γ, $N(u, \epsilon) = N_{\mathcal{S}(\epsilon)}$, $\nu(u, \epsilon)$ and $\overline{\nu}(u, \epsilon)$ are all in the same plane since $\partial \mathcal{S}(\epsilon)$ is a curve on $\mathcal{S}(\epsilon)$ as well as on the container wall Γ.

To get a volume preserving admissible family of comparison surfaces we replace $z_0(u, \epsilon)$ through

$$z^*(u, \epsilon, q) := z_0(u, \epsilon) + q\xi_0(u)N_0(u),$$

where $q \in \mathbb{R}$, $N_0 = N_{\mathcal{S}_0}$, supp $\xi_0 \subset U$, and

$$\int_{\mathcal{S}_0} \xi_0(u) \, dA = 1.$$

Then $\mathcal{S}^*(\epsilon, q)$, defined by $z^*(u, \epsilon, q)$, is a family of admissible comparison surfaces, in general not yet volume preserving, see Fig. 3.15.

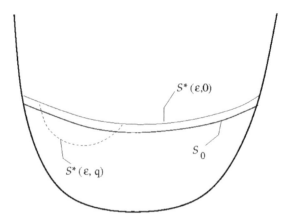

Fig. 3.15   Volume preserving comparison surfaces

We recall that $\mathcal{V}$ is the set of all $\xi N_{\mathcal{S}_0} + \eta T_{\mathcal{S}_0}$, where (ξ, η) satisfies the boundary condition $\xi(u) \cos\gamma - \eta(u) \sin\gamma = 0$ on $\partial \mathcal{S}_0$, and $\mathcal{V}_0$ is the subset of $\mathcal{V}$ such that $\int_{\mathcal{S}_0} \xi\, dA = 0$.

Lemma 3.4. *For given $\zeta \in \mathcal{V}$ satisfying the side condition*

$$\int_{\mathcal{S}_0} \xi\, dA = 0,$$

there exists a regular function $q = q(\epsilon)$ such that the family of admissible configurations given by $\mathcal{S}(\epsilon)$, where $\mathcal{S}(\epsilon)$ is defined through $z(u, \epsilon) = z^*(u, \epsilon, q(\epsilon))$, is volume preserving for all $\epsilon \in \mathbb{R}^2$, $|\epsilon| < \epsilon_0$. The function $q(\epsilon)$ satisfies $q(0) = 0$ and $q'(0) = 0$.

Proof. We have, see Fig. 3.15,

$$\Omega_l(\mathcal{S}^*(\epsilon, q)) = \Omega^1 \cup \Omega^2 \cup \Omega_l(\mathcal{S}_0),$$

where

$$\Omega^1 = \{z^*(u, t, 0) : u \in U,\ 0 < t < \epsilon\},$$
$$\Omega^2 = \{z^*(u, \epsilon, \tau) : u \in U,\ 0 < \tau < q\}.$$

Here we assume for simplicity that $\mathcal{S}^*(\epsilon, q)$ is above of $\mathcal{S}_0$ and of $\mathcal{S}^*(\epsilon, 0)$. The final formula is valid for every $\mathcal{S}^*(\epsilon, q)$, see a remark in [Dierkes, Hildebrandt, Küster and Wohlrab (1992)], p. 81.

Set

$$f(\epsilon, q) = |\Omega_l(\mathcal{S}^*(\epsilon, q))|.$$

From

$$\det \frac{\partial z^*(u,t,0)}{\partial(u,t)} = \langle z_\alpha^* \wedge z_\beta^*, z_t^* \rangle$$

$$= \langle N^*(u,t,0), z_t^*(u,t,0)\rangle W(u,t,0),$$

$$\det \frac{\partial z^*(u,\epsilon,\tau)}{\partial(u,\tau)} = \langle z_\alpha^* \wedge z_\beta^*, z_\tau^* \rangle$$

$$= \langle N^*(u,\epsilon,\tau), z_\tau^*(u,\epsilon,\tau)\rangle W(u,\epsilon,\tau)$$

we find that

$$f(\epsilon,q) = \int_U \int_0^\epsilon \langle N^*(u,t,0), z_t^*(u,t,0)\rangle W(u,t,0)\,dudt$$

$$+ \int_U \int_0^q \langle N^*(u,\epsilon,\tau), z_\tau^*(u,\epsilon,\tau)\rangle W(u,\epsilon,\tau)\,dud\tau$$

$$+ |\Omega_l(\mathcal{S}_0)|.$$

The assertion of the lemma follows since

$$f_q(0,0) = \int_{\mathcal{S}_0} \xi_0\,dA = 1,$$

$$f_\epsilon(0,0) = \int_{\mathcal{S}_0} \xi\,dA.$$

$\square$

Lemma 3.5.

$$\frac{d}{d\epsilon}|\mathcal{S}(\epsilon)| = -2\int_U H(u,\epsilon)\langle N(u,\epsilon), z_\epsilon(u,\epsilon)\rangle W(u,\epsilon)\,du$$

$$+ \int_{\partial U} \langle z_\epsilon(u,\epsilon), \nu(u,\epsilon)\rangle\,ds(\epsilon),$$

where $H(u,\epsilon)$ denotes the mean curvature of $\mathcal{S}(\epsilon)$ at $u \in U$, and

$$ds(\epsilon) = |Z_\tau(\tau,\epsilon)|d\tau, \quad Z(\tau,\epsilon) := z(u(\tau),\epsilon),$$

here $u(\tau)$ is a regular parameter representation of ∂U.

Proof. Set $u = (\alpha, \beta)$, then

$$|\mathcal{S}(\epsilon)| = \int_U \sqrt{E(u,\epsilon)G(u,\epsilon) - F^2(u,\epsilon)}\,du$$

and

$$\frac{d}{d\epsilon}|\mathcal{S}(\epsilon)| = \int_U \frac{1}{W(u,\epsilon)}\big(E(u,\epsilon)\langle z_\beta(u,\epsilon), z_{\beta\epsilon}(u,\epsilon)\rangle$$

$$- F(u,\epsilon)[\langle z_\alpha(u,\epsilon), z_{\beta\epsilon}(u,\epsilon)\rangle + \langle z_\beta(u,\epsilon), z_{\alpha\epsilon}(u,\epsilon)\rangle]$$

$$+ G(u,\epsilon)\langle z_\alpha(u,\epsilon), z_{\alpha\epsilon}(u,\epsilon)\rangle\big)\,du.$$

The formula of the lemma follows by integration by parts, see [Dierkes, Hildebrandt, Küster and Wohlrab (1992); Dierkes, Hildebrandt, and Sauvigny (2010)], p. 45, p. 44, resp., and by using the formula $\triangle z = 2HN$, see [Dierkes, Hildebrandt, Küster and Wohlrab (1992); Dierkes, Hildebrandt, and Sauvigny (2010)], p. 71, p. 72, resp., for a proof of this formula. ☐

Corollary 3.3.

$$\left[\frac{d}{d\epsilon}|\mathcal{S}(\epsilon)|\right]_{\epsilon=0} = -2\int_{S_0} H(u,0)\xi(u)\, dA + \int_{\partial S_0} \eta(u)\, ds.$$

Proof. The corollary follows since

$$z_\epsilon(u,0) = \zeta,$$

and $\zeta = \xi(u)N_{\mathcal{S}_0}(u) + \eta(u)T_{\mathcal{S}_0}$. ☐

Lemma 3.6.

$$\frac{d}{d\epsilon}|\mathcal{W}(\mathcal{S}(\epsilon))| = \int_{\partial U} \langle Z_{0,\epsilon}(\tau,\epsilon), \overline{\nu}(u,\epsilon)\rangle|Z_{0,\tau}(\tau,\epsilon)|\, d\tau,$$

where $Z_0(\tau,t) = z_0(u(\tau),t)$ and $u(\tau)$ is a regular parameter representation of ∂U.

Proof. Let $u(\tau)$ be a regular parameter representation of ∂U. Then

$$|\mathcal{W}(\mathcal{S}(\epsilon))| = \int_{\partial U}\int_0^\epsilon |Z_{0,\tau}(\tau,t) \wedge Z_{0,t}(\tau,t)|\, d\tau dt + |\mathcal{W}(\mathcal{S}(0))|.$$

Here we assume again that $\partial\mathcal{S}(\epsilon)$ is above of $\partial\mathcal{S}(0)$. The final formula is valid for every $\partial_1\mathcal{S}(\epsilon)$, see a remark in [Dierkes, Hildebrandt, Küster and Wohlrab (1992)], p. 81. The assertion of the lemma follows since the tangential plane on the container wall at $P \in \partial\mathcal{S}(t)$ is spanned by the orthogonal vectors $\overline{\nu}(u(\tau),t)$ and $\mathbf{t}(\tau,t) = Z_{0,\tau}(\tau,t)/|Z_{0,\tau}(\tau,t)|$. We recall that $\overline{\nu}(u(\tau),t)$ and $\mathbf{t}(\tau,t)$ are orthogonal vectors. Then $Z_{0,t}(\tau,t) = a\overline{\nu}(u(\tau),t) + b\mathbf{t}(\tau,t)$ with scalar functions a and b depending on τ and t. Using the Lagrange identity, we find that

$$\begin{aligned}
|Z_{0,\tau}(\tau,t) \wedge Z_{0,t}(\tau,t)|^2 &= |Z_{0,\tau}(\tau,t)|^2|Z_{0,t}(\tau,t)|^2 - \langle Z_{0,\tau}(\tau,t), Z_{0,t}(\tau,t)\rangle^2 \\
&= |Z_{0,\tau}(\tau,t)|^2(a^2+b^2) \\
&\quad - \langle Z_{0,\tau}(\tau,t), a\overline{\nu}(u(\tau),t) + b\mathbf{t}(\tau,t)\rangle^2 \\
&= |Z_{0,\tau}(\tau,t)|^2(a^2+b^2) - b^2|Z_{0,\tau}(\tau,t)|^2 \\
&= a^2|Z_{0,\tau}(\tau,t)|^2 \\
&= \langle Z_{0,t}(\tau,t), \overline{\nu}(u(\tau),t)\rangle^2|Z_{0,\tau}(\tau,t)|^2
\end{aligned}$$

☐

Proof. Since $\langle N(\epsilon), \nu(\epsilon)\rangle = 0$ we get

$$\langle N(0), \nu'(0)\rangle = -\langle N'(0), \nu(0)\rangle.$$

From Lemma 3.8 and $\nu_0 = \nu_0^\gamma x_{,\gamma}$ it follows

$$\begin{aligned}
\langle N'(0), \nu_0\rangle &= -g^{\alpha\beta}\langle N(0), \zeta_{,\beta}\rangle\langle x_{,\alpha}, \nu_0^\gamma x_{,\gamma}\rangle \\
&= -g^{\alpha\beta}g_{\alpha\gamma}\langle N(0), (\xi N(0) + \eta^k x_{,k})_{,\beta}\rangle\nu_0^\gamma \\
&= -\langle N(0), (\xi N(0) + \eta^k x_{,k})_{,\beta}\rangle\nu_0^\beta \\
&= -\langle N(0), \xi_{,\beta}N(0) + \xi N(0)_{,\beta} + \eta^k_{,\beta}x_{,k} + \eta^k x_{,k\beta}\rangle\nu_0^\beta \\
&= -\xi_{,\beta}\nu_0^\beta - \langle N(0), \eta^k x_{k\beta}\rangle\nu_0^\beta \\
&= -\frac{\partial\xi}{\partial\nu_0} + \langle\frac{\partial N(0)}{\partial\nu_0}, \nu_0\rangle\eta.
\end{aligned}$$

The previous equation holds since we get from $\langle N(0), x_{,k}\rangle = 0$ that

$$\langle N(0), x_{,k\beta}\rangle = -\langle N(0)_{,\beta}, x_{,k}\rangle.$$

thus, since $\eta^k = \eta\nu_0^k$,

$$\begin{aligned}
\langle N(0), x_{,k\beta}\rangle\eta^k\nu_0^\beta &= -\eta\langle N(0)_{,\beta}, x_{,k}\rangle\nu_0^k\nu_0^\beta \\
&= -\eta\langle\frac{\partial N(0)}{\partial\nu_0}, \nu_0\rangle.
\end{aligned}$$

The assertion of the lemma follows from the formula

$$\langle\frac{\partial N(0)}{\partial\nu_0}, \nu_0\rangle = -K_{\mathcal{S}_0}$$

at $\partial\mathcal{S}_0$. □

For the convenience of the reader we will give the proof of the previous formula.

Lemma 3.10.

$$\langle\frac{\partial N(0)}{\partial\nu_0}, \nu_0\rangle = -K_{\mathcal{S}_0}$$

at $\partial\mathcal{S}_0$.

Proof. Let $P \in \partial\mathcal{S}_0$, and consider the intersection of the plane Π governed by $N(0)$ and ν_0 at P with the surface $\mathcal{S}_0$. Denote by $X(s) = x(u(s))$ the representation of this curve, parameterized by its arc length, and let $\mathcal{N}(s) = N(u(s))$ be the normal on $\mathcal{S}_0$ at $X(s)$. The curvature $K(s)$ of the plane curve $X(s)$ is defined by $K(s) = |X''(s)|$. Since $\langle X'(s), X'(s)\rangle = 1$ it follows that $X''(s)$ is perpendicular on $X'(s)$. Thus $X''(s) = K(s)N_\Pi(s)$, where $N_\Pi(s)$ denotes the normal at $X(s)$ in the plane Π. Since $N_\Pi(s_0) = N(s_0) = N_0$ at $P = X(s_0)$ it follows that

$$X''(s_0) = K_{\mathcal{S}_0} N_0 \qquad (3.48)$$

at $\partial \mathcal{S}_0$. Since at $\partial \mathcal{S}_0$

$$\nu(s) = X'(s) = x_{,\alpha}(u(s))(u^\alpha(s))' = x_{,\alpha}(u(s))\nu^\alpha(s),$$

$$N'(s) = N_{,\alpha}(u(s))(u^\alpha(s))' = N_{,\alpha}(u(s))\nu^\alpha(s) = \frac{\partial N(s)}{\partial \nu(s)}$$

we get

$$\langle \nu, \frac{\partial N(s)}{\partial \nu(s)} \rangle = \langle X'(s), N'(s) \rangle.$$

Hence, since $\langle X'(s), N(s) \rangle = 0$, we find that

$$\langle X''(s_0), N(s_0) \rangle = -\langle \nu_0, \frac{\partial N(0)}{\partial \nu_0} \rangle.$$

The assertion of the lemma follows from formula (3.48). $\qquad \square$

Corollary 3.6. *If $\mathcal{S}_0$ is rotationally symmetric with respect to the x_3-axis, then*

$$\frac{\partial N(0)}{\partial \nu_0} = -K_{\mathcal{S}_0} \nu_0$$

on $\partial \mathcal{S}_0$.

Proof. The vector $\partial N(0)/\partial \nu_0$ is a linear combination of ν_0 and $N(0)$. Since $\langle \partial N(0)/\partial \nu_0, N(0) \rangle = 0$, the formula follows from the lemma above.

$\qquad \square$

Suppose that $\mathcal{S}_0$ is a sufficiently regular surface defined by $x : U \mapsto \mathbb{R}^3$, where $U \subset \mathbb{R}^2$ is a fixed sufficiently regular parameter domain. Let $\mathcal{S}(\epsilon)$ be a perturbed surface given by

$$z(u, \epsilon) = x(u) + \epsilon[\xi(u)N_{\mathcal{S}_0}(u) + \eta^\alpha(u)x_{,\alpha}(u)] + O(\epsilon^2),$$

where ξ, η^α and the remainder are sufficiently regular. The normal $N_{\mathcal{S}_0}$ is defined by $N(u) = (x_{,1} \wedge x_{,2})/|x_{,1} \wedge x_{,2}|$. Let $H(u, \epsilon)$ be the mean curvature of the perturbed surface $\mathcal{S}(\epsilon)$ at $z(u, \epsilon)$. If there is no tangent component $\eta^\alpha x_{,\alpha}$, then we have the well known formula $2H_\epsilon(u, 0) = \triangle \xi + 2(2H^2 - K)\xi$, see [Blaschke (1921)], p. 186, for instance. Here $H = H(u)$ and $K = K(u)$ denote the mean curvature and the Gauss curvature, resp., of the given surface $\mathcal{S}_0$ at $x(u)$. In the general case, a formula for $2H_\epsilon(u, 0)$ till now seems not been mentioned in literature.

Lemma 3.11.

$$2H_\epsilon(u, 0) = \triangle \xi + 2(2H^2 - K)\xi + 2H_{,\alpha}(u, 0)\eta^\alpha.$$

Since

$$b_{\alpha\beta}(\epsilon) = -\langle N_{,\alpha}(\epsilon), z_{,\beta}(\epsilon)\rangle \ ,$$

we have

$$b'_{\alpha\beta}(0) = -\langle N'_{,\alpha}(0), x_{,\beta}\rangle - \langle N_{,\alpha}, z'_{,\beta}(0)\rangle \ .$$

From Lemma 3.8 we have

$$N'(0) = -g^{\alpha\beta}\nu_\beta \, x_{,\alpha} \ .$$

Then

$$\begin{aligned}
\langle N'_{,\alpha}(0), x_{,\beta}\rangle &= -\langle (g^{\gamma\tau}\nu_\tau x_{,\gamma})_{,\alpha}, x_{,\beta}\rangle \\
&= -(g^{\gamma\tau}\nu_\tau)_{,\alpha}g_{\gamma\beta} - g^{\gamma\tau}\nu_\tau\langle x_{,\gamma\alpha}, x_{,\beta}\rangle \\
&= -(g^{\gamma\tau}\nu_\tau)_{,\alpha}g_{\gamma\beta} - g^{\gamma\tau}\nu_\tau\Gamma^\rho_{\gamma\alpha}g_{\rho\beta} \ ,
\end{aligned}$$

where we have used Gauss equations again. Together with

$$\begin{aligned}
\langle N_{,\alpha}, z'_{,\beta}(0)\rangle &= \langle -b^\rho_\alpha x_{,\rho}, \xi^\tau_\beta x_{,\tau} + \nu_\beta N\rangle \\
&= -b^\rho_\alpha \xi^\tau_\beta g_{\rho\tau}
\end{aligned}$$

we obtain formula (3.58) for $b'_{\alpha\beta}(0)$. Combining this formula with formula (3.56), we find, see formulas (3.49) and (3.52) and definitions (3.54), (3.55) of ξ^γ_α and ν_α, that

$$\begin{aligned}
2H'(0) &= (g^{\alpha\tau}\nu_\tau)_{,\alpha} + g^{\gamma\tau}\nu_\tau\Gamma^\alpha_{\gamma\alpha} + b_{\alpha\gamma}\xi^\gamma_\beta g^{\alpha\beta} - b_{\alpha\beta}(\xi^\beta_\gamma g^{\alpha\gamma} + \xi^\alpha_\delta g^{\delta\beta}) \\
&= (g^{\alpha\tau}\nu_\tau)_{,\alpha} + g^{\gamma\tau}\nu_\tau\Gamma^\alpha_{\gamma\alpha} - b_{\alpha\tau}\xi^\alpha_\delta g^{\delta\tau} \\
&= l_1(\lambda) + l_2(\eta) \ ,
\end{aligned}$$

where

$$\begin{aligned}
l_1(\lambda) :&= (g^{\alpha\tau}\lambda_{,\tau})_{,\alpha} + g^{\gamma\tau}\Gamma^\alpha_{\gamma\alpha}\lambda_{,\tau} + b^\gamma_\beta b^\beta_\gamma \lambda, \\
l_2(\eta) :&= (g^{\alpha\tau}b_{\tau\kappa}\eta^\kappa)_{,\alpha} + g^{\gamma\tau}b_{\tau\kappa}\Gamma^\alpha_{\gamma\alpha}\eta^\kappa - b_{\kappa\tau}g^{\delta\tau}\eta^\kappa_{,\delta} - b_{\alpha\tau}g^{\delta\tau}\Gamma^\alpha_{\delta\kappa}\eta^\kappa \\
&= h_\kappa\eta^\kappa \ ,
\end{aligned}$$

with

$$h_\kappa := (g^{\alpha\tau}b_{\tau\kappa})_{,\alpha} + g^{\gamma\tau}\Gamma^\alpha_{\gamma\alpha}b_{\tau\kappa} - b_{\alpha\beta}g^{\delta\beta}\Gamma^\alpha_{\delta\kappa} \ .$$

Since, see [Dierkes, Hildebrandt, Küster and Wohlrab (1992)], p. 55 and p. 19,

$$\Gamma^\alpha_{\gamma\alpha} = \frac{1}{W}W_{,\gamma}$$

and

$$2H = b^1_1 + b^2_2 \ , \qquad K = \det\, (b^\beta_\alpha) \ ,$$

it follows

$$l_1(\lambda) = (g^{\alpha\tau}\lambda_{,\tau})_{,\alpha} + \frac{1}{W}W_{,\gamma}g^{\gamma\tau}\lambda_{,\tau} + 2(2H^2 - K)\lambda$$
$$= \triangle\lambda + 2(2H^2 - K)\lambda\,,$$

see the definition of $\triangle\lambda$ in [Dierkes, Hildebrandt, Küster and Wohlrab (1992)], p. 43. This is the well known formula for the normal variation of $2H$.

Using the formula

$$(g^{\alpha\tau})_{,\kappa} = -g^{\rho\alpha}\Gamma^{\tau}_{\rho\kappa} - g^{\rho\tau}\Gamma^{\alpha}_{\rho\kappa}\,, \tag{3.59}$$

see [Dierkes, Hildebrandt, Küster and Wohlrab (1992)], p. 27, we obtain for the coefficients h_κ in the definition of $l_2(\eta)$

$$h_\kappa = (-g^{\rho\alpha}\Gamma^{\tau}_{\rho\alpha} - g^{\rho\tau}\Gamma^{\alpha}_{\rho\alpha})b_{\tau\kappa} + g^{\alpha\tau}b_{\tau\kappa,\alpha} + g^{\alpha\tau}b_{\tau\kappa}\Gamma^{\tau}_{\gamma\alpha} - b_{\alpha\beta}g^{\delta\beta}\Gamma^{\alpha}_{\delta\kappa}$$
$$= g^{\alpha\tau}b_{\tau\kappa,\alpha} - g^{\rho\alpha}b_{\tau\kappa}\Gamma^{\tau}_{\rho\alpha} - b_{\alpha\beta}g^{\delta\beta}\Gamma^{\alpha}_{\delta\kappa}\,.$$

From the Codazzi equations, see [Dierkes, Hildebrandt, Küster and Wohlrab (1992)], p. 29, we get

$$b_{\tau\kappa,\alpha} = b_{\tau\alpha,\kappa} + \Gamma^{\epsilon}_{\tau\alpha}b_{\epsilon\kappa} - \Gamma^{\epsilon}_{\tau\kappa}b_{\epsilon\alpha},$$

where we have used the symmetry properties $b_{\tau\kappa} = b_{\kappa\tau}$ and $\Gamma^{\epsilon}_{\tau\kappa} = \Gamma^{\epsilon}_{\kappa\tau}$. With formula (3.59) we find finally

$$h_\kappa = g^{\alpha\tau}b_{\alpha\tau,\kappa} + g^{\alpha\tau}(\Gamma^{\epsilon}_{\tau\alpha}b_{\epsilon\kappa} - \Gamma^{\epsilon}_{\tau\kappa}b_{\epsilon\alpha}) - g^{\rho\alpha}b_{\tau\kappa}\Gamma^{\tau}_{\rho\alpha} - b_{\alpha\beta}g^{\delta\beta}\Gamma^{\alpha}_{\delta\kappa}$$
$$= g^{\alpha\tau}b_{\alpha\tau,\kappa} - g^{\alpha\tau}b_{\epsilon\alpha}\Gamma^{\epsilon}_{\tau\kappa} - b_{\alpha\beta}g^{\delta\beta}\Gamma^{\alpha}_{\delta\kappa}$$
$$= (g^{\alpha\tau}b_{\alpha\tau})_{,\kappa} - b_{\alpha\tau}(g^{\alpha\tau})_{,\kappa} - g^{\alpha\tau}b_{\epsilon\alpha}\Gamma^{\epsilon}_{\tau\kappa} - b_{\alpha\beta}g^{\delta\beta}\Gamma^{\alpha}_{\delta\kappa}$$
$$= (g^{\alpha\tau}b_{\alpha\tau})_{,\kappa}$$
$$= 2H_{,\kappa}.$$

$\square$

We assume that admissible comparison surfaces are defined through

$$x(u) + \epsilon\zeta(u) + \frac{\epsilon^2}{2}|\zeta(u)|^2 r(u) + O(\epsilon^3),$$

where $\zeta(u) = \xi(u)N_0(u) + \eta(u)T_{\mathcal{S}_0}(u)$, for all $|\epsilon| < \epsilon_0$, see Sec. 3.4 for the definition of $r(u)$. We recall that we obtained, using such a family, from the equilibrium condition the boundary condition $\xi(u)\cos\gamma - \eta(u)\sin\gamma = 0$ on ∂U, see the proof of Theorem 3.1. We denote by $N(\epsilon)$, $\overline{N}(\epsilon)$ the normals on $\mathcal{S}(\epsilon)$ and Γ, resp., at $\partial\mathcal{S}(\epsilon)$, see Fig. 3.14.

In the following lemma we have $\langle \bar{\nu}, \zeta/|\zeta| \rangle = \pm 1$ since $\bar{\nu} = \pm \zeta/|\zeta|$. If $\zeta = 0$, then we define $\langle \bar{\nu}, \zeta/|\zeta| \rangle$ by a fixed constant, say by 0.

Lemma 3.12.
$$\left[\frac{d}{d\epsilon} \langle N(\epsilon), \overline{N}(\epsilon) \rangle \right]_{\epsilon=0} = \frac{\partial \xi}{\partial \nu} \sin \gamma + K_{\mathcal{S}} \eta \, \sin \gamma - K_\Gamma |\zeta| \langle \bar{\nu}, \zeta/|\zeta| \rangle \sin \gamma.$$

Proof. From Lemma 3.8 it follows that
$$\langle N'(0), \overline{N}(0) \rangle = -g^{\alpha\beta} \langle N(0), \zeta_{,\alpha} \rangle \langle x_{,\alpha}, \overline{N}(0) \rangle.$$

Since
$$\overline{N}(0) = N(0) \cos \gamma - \nu \, \sin \gamma,$$

we have
$$\begin{aligned}
\langle x_{,\alpha}, \overline{N}(0) \rangle &= -\langle x_{,\alpha}, \nu \rangle \sin \gamma \\
&= \langle x_{,\alpha}, x_{,\beta} \rangle \nu^\beta \sin \gamma \\
&= -g_{\alpha\tau} \nu^\tau \sin \gamma.
\end{aligned}$$

Then on $\mathcal{S}_0$,
$$\begin{aligned}
\langle N'(0), \overline{N}(0) \rangle &= g^{\alpha\beta} \langle N(0), \zeta_{,\beta} \rangle g_{\alpha\tau} \nu^\tau \sin \gamma \\
&= \langle N(0), \zeta_{,\beta} \rangle \nu^\beta \sin \gamma \\
&= \xi_{,\beta} \nu^\beta \sin \gamma + \langle N(0), T_{,\beta} \rangle \eta \, \nu^\beta \sin \gamma \\
&= \frac{\partial \xi}{\partial \nu} \sin \gamma - \langle N_{,\beta}(0), T \rangle \eta \, \nu^\beta \sin \gamma \\
&= \frac{\partial \xi}{\partial \nu} \sin \gamma - \langle \frac{\partial N}{\partial \nu}, \nu \rangle \eta \, \sin \gamma \\
&= \frac{\partial \xi}{\partial \nu} \sin \gamma + K_{\mathcal{S}} \eta \, \sin \gamma.
\end{aligned}$$

Next we show that
$$\langle N(0), \overline{N}'(0) \rangle = -K_\Gamma |\zeta| \langle \bar{\nu}, \zeta/|\zeta| \rangle \sin \gamma.$$

A parameter representation of a neighborhood on Γ of $\partial \mathcal{S}_0$ is given by
$$z(\epsilon, \tau) = x(u(\tau)) + \epsilon \zeta(u(\tau)) + \frac{\epsilon^2}{2} |\zeta(u)|^2 r(u(\tau)) + O(\epsilon^3),$$

where $u(\tau)$ is a parameter representation of ∂U, and $|\epsilon| < \epsilon_0$. Set $z_{,1} = z_{,\epsilon}$, $z_{,2} = z_{,\tau}$, then
$$\begin{aligned}
\overline{N}'(0) &= c^\alpha z_{,\alpha}(0, \tau) + c\overline{N}(0) \\
&= c^\alpha z_{,\alpha}(0, \tau), \tag{3.60}
\end{aligned}$$

since $\langle \overline{N}(\epsilon), \overline{N}(\epsilon) \rangle = 1$. Consequently, we have

$$\langle \overline{N}'(0), z_{,\beta}(0, \tau) \rangle = c^\alpha \overline{g}_{\alpha\beta}(0, \tau),$$

where a bar indicates quantities which are related to Γ. Since

$$\langle \overline{N}(\epsilon), z_{,\beta}(\epsilon, \tau) \rangle = 0,$$

$\beta = 1, 2$, it follows

$$\langle \overline{N}'(0), z_{,\beta}(0, \tau) \rangle + \langle \overline{N}(0), z_{,\beta\epsilon}(0, \tau) \rangle = 0$$

Thus

$$c^\alpha = -\overline{g}^{\alpha\beta}(0, \tau) \langle \overline{N}(0), z_{,\beta\epsilon}(0, \tau) \rangle.$$

Then, see formula (3.60),

$$\overline{N}'(0) = -\overline{g}^{\alpha\beta}(0, \tau) \langle \overline{N}(0), z_{,\beta\epsilon}(0, \tau) \rangle z_{,\alpha}(0, \tau).$$

Combining this equation with

$$\langle z_{,\alpha}(0, \tau), N(0) \rangle = \langle z_{,\alpha}(0, \tau), \overline{N}(0) \cos\gamma + \overline{\nu} \sin\gamma \rangle$$
$$= \langle z_{,\alpha}(0, \tau), \overline{\nu} \rangle \sin\gamma$$
$$= \langle z_{,\alpha}(0, \tau), z_{,\rho}(0, \tau) \rangle \overline{\nu}^\rho \sin\gamma$$
$$= \overline{g}_{\alpha\rho} \overline{\nu}^\rho \sin\gamma,$$

we get finally

$$\langle \overline{N}'(0), N(0) \rangle = -\langle \overline{N}(0), z_{,\beta\epsilon}(0, \tau) \rangle \overline{\nu}^\beta \sin\gamma$$
$$= -\langle \overline{N}(0), |\zeta(u(\tau))|^2 r(u(\tau)) \rangle \overline{\nu}^1 \sin\gamma$$
$$\quad - \langle \overline{N}(0), \zeta(u(\tau))_{,\tau} \rangle \overline{\nu}^2 \sin\gamma$$
$$= -\langle \overline{N}(0), |\zeta(u)|^2 K_\Gamma \overline{N}(0) \rangle \overline{\nu}^1 \sin\gamma$$
$$= -K_\Gamma |\zeta|^2 \overline{\nu}^1 \sin\gamma$$
$$= -K_\Gamma |\zeta| \langle \overline{\nu}, \zeta/|\zeta| \rangle \sin\gamma.$$

Here we have used that

$$\overline{\nu} = \pm \frac{\zeta}{|\zeta|} = \overline{\nu}^1 \zeta \equiv \overline{\nu}^1 z_{,1}(0, \tau),$$

which implies that $\overline{\nu}^2 = 0$.   □

Corollary 3.7. *If γ satisfies $0 < \gamma < \pi$, then*

$$\left[\frac{d}{d\epsilon} \langle N(\epsilon), \overline{N}(\epsilon) \rangle \right]_{\epsilon=0} = \left[\frac{\partial \xi}{\partial \nu} + \left(\frac{\cos\gamma}{\sin\gamma} K_S - \frac{1}{\sin\gamma} K_\Gamma \right) \xi \right] \sin\gamma.$$

Proof. We have on ∂U

$$|\zeta|^2 = \xi^2 + \eta^2$$

and, see (3.3),

$$\eta = \frac{\cos\gamma}{\sin\gamma} \xi.$$

□

3.7 Problems

(1) Show that there is a correction term $r(\alpha, \beta; \epsilon)$ such that $\partial \mathcal{S}(\epsilon) \subset \partial \Omega_s$.

(2) Show that N_Γ, $N_\mathcal{S}$ and $T_\mathcal{S}$ are in a common plane, provided that the container wall is sufficiently regular. *Hint:* Denote by g the tangent at $\partial \mathcal{S}$ in $P \in \partial \mathcal{S}$, then all vectors N_Γ, $N_\mathcal{S}$ and $T_\mathcal{S}$ are perpendicular on g.

(3) Show that

$$\det(x_\alpha, x_\beta, N_\mathcal{S}) = \sqrt{E_0 G_0 - F_0^2}.$$

Hint:

$$\det(x_\alpha, x_\beta, \; N_\mathcal{S}) = \sum_{i=1}^{3} N_i D_i,$$

where

$$x_\alpha \wedge x_\beta = (D_1, D_2, D_3), \quad N_\mathcal{S} = (N_1, N_2, N_3).$$

Then

$$\sum_{i=1}^{3} N_i D_1 = \sqrt{D_1^2 + D_2^2 + D_3^2} = |x_\alpha \wedge x_\beta|.$$

Finally use the Lagrange identity

$$(a \wedge b) \cdot (c \wedge d) = (a \cdot c)(b \cdot d) - (b \cdot c)(a \cdot d).$$

(4) Let $\mathcal{S}$ be given by $x_3 = u(x_1, x_2)$. Show that $2H = \operatorname{div} \; Tu$.

(5) Suppose that $\mathcal{S}$ is a capillary surface over $\Omega \subset \mathbb{R}^2$. Show that $N_\mathcal{S} \cdot N_\Gamma = \nu \cdot Tu = \cos \gamma$, where $\Gamma = \partial \Omega \times (-\infty, \infty)$.

(6) Derive the final formula for the ascent of a liquid at a vertical wall, see Sec. 3.3.1.

(7) Let Ω be the half plane $\Omega = \{(x_1, x_2) : x_1 > 0\}$. Suppose that $u \in C^2(\Omega)$ satisfies $\operatorname{div} \; Tu = \lambda$, where λ is constant. Prove that $\lambda = 0$, i. e., u is a solution of the minimal surface equation. The same result holds if Ω is a sector in the plane.

(8) Let $u(x) = v(r)$ be a rotationally symmetric surface with respect to the x_3-axis, $r = \sqrt{x_1^2 + x_2^2}$. Show that

$$\operatorname{div} \; Tu = \frac{1}{r} \left(\frac{r v'(r)}{\sqrt{1 + v'(r)^2}} \right)'.$$

(9) See the problem of the ascent of liquid at two parallel vertical plates. Prove the asymptotic formula for the constant C.

(10) See the problem of the ascent of liquid at two parallel vertical plates. Consider the case of different (constant) contact angles on each plate and express the solution of the capillary problem through elliptic integrals.

(11) See the problem of the ascent of liquid at two vertical walls. Find asymptotic formulas for the heights u_0, u_1 which are uniform in $0 \leq \gamma \leq \pi/2$.

(12) See the problem of the ascent of liquid at two vertical walls. Find asymptotic formulas in the case of different contact angles at each plate.

(13) Show that there exists a zero gravity capillary interface between two coaxial cylinders for any given contact angles γ_1, $\gamma_2 \in [0, \pi]$, see Sec. 3.3
Hint: Show that $|f(s)| \leq 1$.

(14) Consider a cross section as shown in Fig. 3.16. Show that there exist a solution of the zero gravity problem if $1 \leq a < 1.95$ holds.

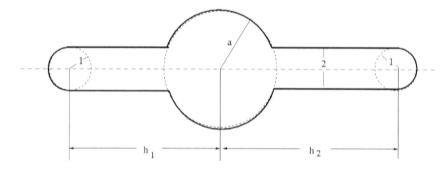

Fig. 3.16   Slit with a bore

(15) Consider a cross section as sketched in Fig. 3.17 and find a, r_1 and r_2 such there exists a zero gravity solution for every contact angle.

(16) See Fig. 3.5. Find a lower bound V_0 such that there is no dry spot on the bottom or that $\mathcal{S}^+$ is above of $\mathcal{S}^-$, resp., if $V > V_0$.

(17) Let $0 < \alpha \leq \pi$. Set

$$\Omega_\alpha = \{(r, \theta) : r > 0, \ 0 < |\theta| \leq \pi\},$$

where (r, θ) are polar coordinates with center at the origin and consider the linear boundary value problem $\triangle u = 0$ in Ω_α, $\partial u / \partial \nu = \cos \gamma$, where $0 \leq \gamma < 2\pi$. Show that there is a solution $u \in C^2(\overline{\Omega_\alpha} \setminus \{0\})$. (Thus, there is a striking difference between the original problem (3.22), (3.23) and the linearized problem.)

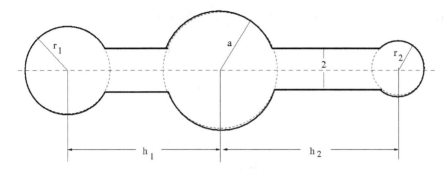

Fig. 3.17   Slit with bores

(18) Set

$$D(\epsilon) = \langle z'(\epsilon), N(\epsilon) \rangle W(\epsilon),$$

where

$$z(\epsilon) = x(u) + \epsilon[\xi(u)N_{\mathcal{S}_0}(u) + \eta^\alpha x_{,\alpha}(u)] + O(\epsilon^2)$$

and the remainder is sufficiently regular. Show that $D(0) = \xi W(0)$.

(19) Consider the same family of surfaces $z(\epsilon)$ as in the previous exercise. Set

$$W(\epsilon) = \sqrt{E(\epsilon)G(\epsilon) - F^2(\epsilon)} \equiv \det\left(g_{\alpha\beta}(\epsilon)\right).$$

Show that

$$W'(0) = -2HW(0)\xi + (\eta^\alpha W(0))_{,\alpha}.$$

Hint. Use Gauss and Weingarten equations, see [Dierkes, Hildebrandt, Küster and Wohlrab (1992)], p. 55.

(20) Consider the family of surfaces of second order defined at the beginning of Sec. 10.2. Find a formula for $D'(0)$.

(21) Consider pure normal perturbations defined by

$$z(\epsilon) = x(u) + \epsilon\xi(u)N(u).$$

Recover the formula

$$\left[\frac{d^2}{d\epsilon^2}|\mathcal{S}(\epsilon)|\right]_{\epsilon=0} = \int_{\mathcal{S}_0} \left(2K_0\xi^2 + |\nabla\xi|^2\right)\, dA,$$

see [Blaschke (1921)], p. 184, from Lemma 3.5 *Hint.*

$$2H_\epsilon(u, 0) = \triangle\xi + 2(2H(u, 0) - K(u, 0))\xi,$$
$$D_\epsilon(u, 0) = -2H(u, 0)\xi^2 W(u, 0),$$
$$\langle N(u, 0), \nu_\epsilon(u, 0) \rangle = \frac{\partial\xi}{\partial\nu_0}.$$

(22) Consider pure normal perturbations defined by

$$z(\epsilon) = x(u) + \epsilon\xi(u)N(u).$$

Set

$$D(\epsilon) = \langle z'(\epsilon), N(\epsilon)\rangle W(\epsilon).$$

Show that

$$D(\epsilon) = \det(z_\alpha, z_\beta, z_\epsilon)$$

and, by using the following hint,

$$D'(0) = -2H\zeta^2\sqrt{E_0G_0 - F_0^2}.$$

Hint: Consider $(D(\epsilon))^2$, where

~~$D(\epsilon) \; \det(z_\alpha, z_\beta, z_\epsilon),$~~

and use the formula

$$(D(\epsilon))^2 = \det\begin{pmatrix} z_\alpha \cdot z_\alpha & z_\alpha \cdot z_\beta & z_\alpha \cdot z_\epsilon \\ z_\beta \cdot z_\alpha & z_\beta \cdot z_\beta & z_\beta \cdot z_\epsilon \\ z_\epsilon \cdot z_\alpha & z_\epsilon \cdot z_\beta & z_\epsilon \cdot z_\epsilon \end{pmatrix}.$$

Chapter 4

Floating drops

4.1 Governing energy

Consider a liquid drop Ω_2 floating on another liquid Ω_1. For simplicity we consider the case that the liquid above of $\Omega_1 \cup \Omega_2$ is vapor Ω_3 of density zero, see Fig. 4.1 where we followed, in principle, the notations of [Slobozhanin (1986)]. The vectors ν_k in the above Fig. 4.1 denote tangent vectors in $\mathcal{S}_k$

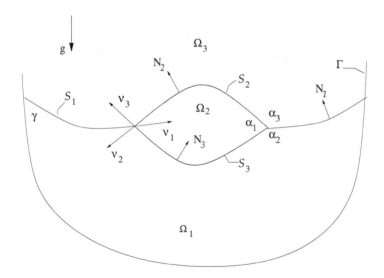

Fig. 4.1   Floating drop, notations

which are perpendicular on the edge of the drop. Set $\mathcal{S} = \mathcal{S}_1 \cup \mathcal{S}_2 \cup \mathcal{S}_3$. Since we assume that the volumes of Ω_1 and Ω_2 are fixed, we have the side conditions

$$|\Omega_1(\mathcal{S})| = C_1, \quad |\Omega_2(\mathcal{S})| = C_2. \tag{4.1}$$

Suppose that the energy of the problem is given by

$$\mathcal{E}(\mathcal{S}) = \sum_{k=1}^{3} \sigma_k |\mathcal{S}_k| - \sigma_1 \beta |\mathcal{W}(\mathcal{S})| \tag{4.2}$$

$$+ \int_{\Omega_1(\mathcal{S})} F_1(x) \, dx + \int_{\Omega_2(\mathcal{S})} F_2(x) \, dx,$$

where

σ_k surface tension of $\mathcal{S}_k$ (positive constants),

β (relative) adhesion coefficient between the fluid Ω_1 and the container wall Γ,

$\mathcal{W}(\mathcal{S})$ wetted part of the container,

$F_1 = g\rho_1 x_3$, where ρ_1 is the density of the liquid occupied by Ω_1,

$F_2 = g\rho_2 x_3$, where ρ_2 is the density of the liquid occupied by Ω_2.

We suppose that β, ρ_1 and ρ_2 are positive constants.

4.2 Equilibrium conditions

Let $\mathcal{S}_1$ be given by $x_1(u)$, $\mathcal{S}_2$ by $x_2(v)$ and $\mathcal{S}_3$ by $x_3(w)$, where $u \in \mathbf{U}$, $v \in \mathbf{V}$, $w \in \mathbf{W}$ are parameter domains in $\mathbb{R}^2$, $\mathbf{U}$ is an annulus and $\mathbf{V}$, $\mathbf{W}$ are disks. We choose the parameter representation of $x_1(u)$, $x_2(v)$, $x_3(w)$, resp., such that the normals $N_k := x_{k,1} \wedge x_{k,2} / |x_{k,1} \wedge x_{k,2}|$ are oriented as shown in Fig. 4.1.

Suppose that the edge of the drop Λ, i. e., the intersection of the closed surfaces $\mathcal{S}_k$, is defined by

$$x_1(u(\tau)) = x_2(v(\tau)) = x_3(w(\tau)),$$

where $u(\tau)$ is a regular parameter representation of $\partial_2 \mathbf{U}$, $v(\tau)$ a regular parameter representation of $\partial \mathbf{V}$ and $w(\tau)$ a regular parameter representation of $\partial \mathbf{W}$.

Let $\nu(\tau) \equiv \nu(x(u(\tau)))$ be a given sufficiently regular vector field, see Fig. 4.2, and the perturbed configuration $(\mathcal{S}_1(\epsilon), \mathcal{S}_2(\epsilon), \mathcal{S}_3(\epsilon))$, see Fig. 4.2, is defined as follows. The perturbed surface $\mathcal{S}_1(\epsilon)$ is given by

$$z_1(u, \epsilon) = x_1(u) + \epsilon \zeta_1(u) + r(u, \epsilon),$$

where

$$\zeta_1(u) = \xi_1(u) N_1(u) + \eta_1^\alpha(u) x_{1,\alpha}(u)$$

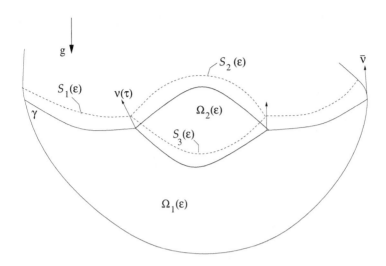

Fig. 4.2   Perturbed configuration, notations

with $\zeta_1 = a(\tau)\nu(\tau)$ on Λ and $\zeta_1 = b(\tau)\bar{\nu}(\tau)$ on $\partial_1 \mathbf{U}$, i. e., on the intersection of the container wall with $\mathcal{S}_1$. The functions $a(\tau)$ and $b(\tau)$ are given sufficiently regular scalar functions. The remainder $r(u, \epsilon)$ satisfies $r(u, \epsilon) = O(\epsilon)$ in a neighborhood of $\partial_1 \mathbf{U}$ and $r = 0$ in a neighborhood of $\partial_2 \mathbf{U}$.

Further $\mathcal{S}_2$ is defined through

$$z_2(v, \epsilon) = x_2(v) + \epsilon\zeta_2(v),$$

where

$$\zeta_2(v) = \xi_2(v)N_2(v) + \eta_2^\alpha(v)x_{2,\alpha}(v)$$

with $\zeta_2 = a(\tau)\nu(\tau)$ on Λ.

Finally $\mathcal{S}_3$ is given by

$$z_3(w, \epsilon) = x_3(w) + \epsilon\zeta_3(w),$$

where

$$\zeta_3(w) = \xi_3(w)N_3(w) + \eta_3^\alpha(w)x_{3,\alpha}(w)$$

with $\zeta_3 = a(\tau)\nu(\tau)$ on Λ.

Let $\mathcal{V}$ be the set of all $\zeta = (\zeta_1, \zeta_2, \zeta_3)$ satisfying $\zeta_1 = b(\tau)\bar{\nu}(\tau)$ on $\partial_1 \mathbf{U}$ and $\zeta_k = b(\tau)\nu(\tau)$, $k = 1, 2, 3$, on Λ. The subset $\mathcal{V}_0$ is defined by

$$\mathcal{V}_0 = \{\zeta \in \mathcal{V} : \int_{\mathcal{S}_1} \xi_1 \, dA + \int_{\mathcal{S}_3} \xi_3 \, dA = 0 \text{ and } \int_{\mathcal{S}_2} \xi_2 \, dA - \int_{\mathcal{S}_3} \xi_3 \, dA = 0\}.$$

Following the reasoning of the proof of Lemma 3.4 of the previous chapter, we find that for given $\zeta \in \mathcal{V}_0$ there exists a volume preserving family of comparison configurations defined through

$$\mathcal{S}^*(\epsilon) = (\mathcal{S}_1^*(\epsilon), \mathcal{S}(\epsilon), \mathcal{S}_3(\epsilon)),$$

where $\mathcal{S}_1^*(\epsilon)$ is given by

$$z_1^*(u, \epsilon, q_1(\epsilon), q_2(\epsilon)) = z_1(u, \epsilon) + q_1(\epsilon)h_1(u)N_1(u) + q_2(\epsilon)h_2(u)N_1(u).$$

Here h_1, h_2 are fixed sufficiently regular scalar functions satisfying

$$\text{supp } h_1, \ \text{supp } h_2 \subset \mathbf{U}, \ \text{supp } h_1 \cap \text{supp } h_2 = \emptyset$$

and

$$\int_{\mathcal{S}_1} h_k \, dA = 1, \ k = 1, 2.$$

The functions $q_k(\epsilon)$ satisfy $q_k(0) = q_k'(0) = 0$. Volume preserving means here that

$$|\Omega_1(\mathcal{S}^*(\epsilon))| = C_1, \quad |\Omega_2(\mathcal{S}^*(\epsilon))| = C_2,$$

where C_1 is the volume of the liquid occupied by Ω_1 and C_2 is the volume of the liquid occupied by Ω_2, see Fig. 4.1.

Define the Lagrange function

$$\mathcal{L}(\mathcal{S}, \lambda) = \mathcal{E}(\mathcal{S}) + \lambda^1(|\Omega_1(\mathcal{S})| - C_1) + \lambda^2(|\Omega_2(\mathcal{S})| - C_2).$$

From a Lagrange multiplier rule, see Chap. 10, it follows that there exists a real $\lambda_0 = (\lambda_0^1, \lambda_0^2)$ such that

$$\left[\frac{d}{d\epsilon}\mathcal{L}(\mathcal{S}(\epsilon), \lambda_0)\right]_{\epsilon=0} = 0,$$

provided that

$$\mathcal{E}(\mathcal{S}^*(\epsilon)) \geq \mathcal{E}(\mathcal{S}(0))$$

holds for all $|\epsilon| \leq \epsilon_0$.

Formulas (4.10), (4.11) from Lemma 4.1, Lemma 3.5 and the corollary to Lemma 3.6 from the appendix to the previous chapter we get the formula

$$
\left[\frac{d}{d\epsilon}\mathcal{L}(\mathcal{S}(\epsilon),\lambda_0)\right]_{\epsilon=0} = -2\sum_{k=1}^{3}\sigma_k\int_{\mathcal{S}_k(0)} H_k\xi_k \, dA
$$

$$
+ \sum_{k=1}^{3}\sigma_k\int_{\Lambda} a\langle\nu,\nu_k\rangle \, ds
$$

$$
+ \sigma_1\int_{\partial_1\mathcal{S}_1(0)} b\left(\langle\overline{\nu},\nu_1\rangle - \beta\right) \, ds
$$

$$
+ \int_{\mathcal{S}_1(0)} F_1(x_1(u))\xi_1(u) \, dA + \int_{\mathcal{S}_3(0)} F_1(x_3(w))\xi_3(w) \, dA
$$

$$
+ \int_{\mathcal{S}_2(0)} F_2(x_2(v))\xi_2(v) \, dA - \int_{\mathcal{S}_3(0)} F_2(x_3(w))\xi_3(w) \, dA
$$

$$
+ \lambda_0^1\left(\int_{\mathcal{S}_1(0)} \xi_1(u) \, dA + \int_{\mathcal{S}_3(0)} \xi_3(w) \, dA\right)
$$

$$
+ \lambda_0^2\left(\int_{\mathcal{S}_2(0)} \xi_2(v) \, dA - \int_{\mathcal{S}_3(0)} \xi_3(w) \, dA\right).
$$

Theorem 4.1. *The equation*

$$
\left[\frac{d}{d\epsilon}\mathcal{L}(\mathcal{S}(\epsilon),\lambda_0)\right]_{\epsilon=0} = 0
$$

implies the following equilibrium conditions

$$
-2\sigma_1 H(u) + F_1(u) + \lambda_0^1 = 0 \quad on \ \mathcal{S}_1, \tag{4.3}
$$

$$
-2\sigma_2 H(v) + F_2(v) + \lambda_0^2 = 0 \quad on \ \mathcal{S}_2, \tag{4.4}
$$

$$
-2\sigma_3 H(w) + F_1(w) - F_2(w) + \lambda_0^1 - \lambda_0^2 = 0 \quad on \ \mathcal{S}_3, \tag{4.5}
$$

$$
\cos\gamma = \beta \quad on \ \partial_1\mathcal{S}_1, \tag{4.6}
$$

$$
\sum_{k=1}^{3}\sigma_k\nu_k = 0 \quad on \ \Lambda. \tag{4.7}
$$

Proof. As in the proof of the equilibrium conditions for a capillary interface we find equations (4.3)–(4.5). Concerning equation (4.6) we recall that $\langle\overline{\nu},\nu_1\rangle = \cos\gamma$, see Fig. 4.3. Since ν is an arbitrary vector perpendicular on the edge Λ of the drop we get equation (4.7). $\qquad\square$

Remark 4.1. The boundary condition (4.7) was derived by [Neumann (1894)], p. 161.

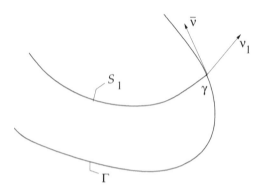

Fig. 4.3 $\langle \bar{\nu}, \nu_1 \rangle = \cos \gamma$

From formula (4.7) we find the following corollaries, see also [Slobozhanin (1986)], p. 482.

Corollary 4.1. *Let* $i \neq j$, $i \neq k$, $j \neq k$, *then*

$$\sigma_i \leq \sigma_j + \sigma_k.$$

Corollary 4.2. *Let* $i \neq j$, $i \neq k$, $j \neq k$, *then*

$$\cos \alpha_i = \frac{\sigma_i^2 - \sigma_j^2 - \sigma_k^2}{2\sigma_j \sigma_k}. \qquad (4.8)$$

Remark 4.2. The inequalities of the first corollary show that a drop is not in an equilibrium if the inequalities

$$\sigma_i < |\sigma_j - \sigma_k|$$

hold, where $i \neq j$, $i \neq k$, $j \neq k$. The previous inequalities were discovered by [Marangoni (1871)] from experiments with fluids, in particular on the water surface of a container with diameter of about 70 meters in the Tuileries in Paris.

Remark 4.3. Suppose that $\alpha_2 = \pi$, see Fig. 4.4. Then we find from the above formula (4.8) for $\cos \alpha_i$ that

$$\cos \alpha_1 = \frac{\sigma_1 - \sigma_3}{\sigma_2}. \qquad (4.9)$$

Replacing the liquid which occupies Ω_1 through a solid material, then the angle α_1 defined through formula (4.9) is called *Young's angle*. In fact, this formula fails in general for liquid/solid interfaces. This was a discovery of [Finn (2010)]. The problem is here to find the surface tensions σ_1 and σ_3.

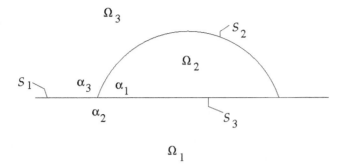

Fig. 4.4 Drop on a horizontal plane

4.3 Appendix

Suppose that the domain $\Omega_1(\epsilon) := \Omega_1(\mathcal{S}(\epsilon))$ is given by
$$\Omega_1(\epsilon) = \Omega_1(0) \cup \Omega^1(\epsilon) \cup \Omega^3(\epsilon),$$
where
$$\Omega^1(\epsilon) = \{z_1(u,t) : \ u \in \mathbf{U}, \ 0 < t < \epsilon\},$$
$$\Omega^3(\epsilon) = \{z_3(w,t) : \ w \in \mathbf{W}, \ 0 < t < \epsilon\},$$
and that $\Omega_2(\epsilon) = \Omega_2(\mathcal{S}(\epsilon))$ is defined by
$$\Omega_2(\epsilon) = \big(\Omega_2(0) \cup \Omega^2(\epsilon)\big) \setminus \Omega^3(\epsilon),$$
where
$$\Omega^2(\epsilon) = \{z_2(v,t) : \ v \in \mathbf{V}, \ 0 < t < \epsilon\},$$
see Fig. 4.2. We assume here that the surfaces $\mathcal{S}_k(\epsilon)$ are above of $\mathcal{S}_k(0)$. The final formulas are valid for the general case, see a remark in [Dierkes, Hildebrandt, Küster and Wohlrab (1992)], p. 81. Then
$$\int_{\Omega_1(\epsilon)} F_1(x)\,dx = \int_{\Omega_1(0)} F_1(x)\,dx + \int_{\Omega^1(\epsilon)} F_1(x)\,dx + \int_{\Omega^3(\epsilon)} F_1(x)\,dx,$$
where
$$\int_{\Omega^1(\epsilon)} F_1(x)\,dx = \int_{\mathbf{U}} \int_0^\epsilon F_1(z_1(u,t)) \det \frac{\partial z_1(u,t)}{\partial(u,t)}\,du\,dt$$
$$= \int_{\mathbf{U}} \int_0^\epsilon F_1(z_1(u,t)) \langle N_1(u,t), z_{1,t}(u,t)\rangle W(u,t)\,du\,dt,$$
$$\int_{\Omega^3(\epsilon)} F_1(x)\,dx = \int_{\mathbf{W}} \int_0^\epsilon F_1(z_3(w,t)) \det \frac{\partial z_3(w,t)}{\partial(w,t)}\,du\,dt$$
$$= \int_{\mathbf{W}} \int_0^\epsilon F_1(z_3(w,t)) \langle N_3(w,t), z_{3,t}(w,t)\rangle W(w,t)\,du\,dt,$$

and

$$\int_{\Omega_2(\epsilon)} F_2(x)\, dx = \int_{\Omega_2(0)} F_2(x)\, dx + \int_{\Omega^2(\epsilon)} F_2(x)\, dx - \int_{\Omega^3(\epsilon)} F_2(x),$$

where

$$\int_{\Omega^2(\epsilon)} F_2(x)\, dx = \int_{\mathbf{V}} \int_0^\epsilon F_2(z_2(v,t)) \det \frac{\partial z_2(v,t)}{\partial(v,t)}\, dv dt$$

$$= \int_{\mathbf{V}} \int_0^\epsilon F_2(z_2(v,t))\langle N_2(v,t), z_{2,t}(v,t)\rangle W(v,t)\, dv dt,$$

$$\int_{\Omega^3(\epsilon)} F_2(x)\, dx = \int_{\mathbf{W}} \int_0^\epsilon F_2(z_3(w,t))\langle N_3(w,t), z_{3,t}(w,t)\rangle W(w,t)\, du dt.$$

Thus we get

Lemma 4.1.

$$\frac{d}{d\epsilon} \int_{\Omega_1(\epsilon)} F_1(x) dx = \int_{\mathbf{U}} F_1(z_1(u,\epsilon))\langle N_1(u,\epsilon), z_{1,\epsilon}\rangle W(u,\epsilon) du \qquad (4.10)$$

$$+ \int_{\mathbf{W}} F_1(z_3(w,\epsilon))\langle N_3(w,\epsilon), z_{3,\epsilon}\rangle W(w,\epsilon) dw,$$

$$\frac{d}{d\epsilon} \int_{\Omega_2(\epsilon)} F_2(x) dx = \int_{\mathbf{V}} F_1(z_2(v,\epsilon))\langle N_2(v,\epsilon), z_{2,\epsilon}\rangle W(v,\epsilon) dv \qquad (4.11)$$

$$- \int_{\mathbf{W}} F_2(z_3(w,\epsilon))\langle N_3(w,\epsilon), z_{3,\epsilon}\rangle W(w,\epsilon) dw.$$

4.4   Problems

(1) Let $i \neq j$, $i \neq k$, $j \neq k$. Then

$$\sigma_i \leq \sigma_j + \sigma_k.$$

(2) Let $i \neq j$, $i \neq k$, $j \neq k$. Then

$$\cos \alpha_i = \frac{\sigma_i^2 - \sigma_j^2 - \sigma_k^2}{2\sigma_j \sigma_k}.$$

(3) Find equilibrium configurations in the zero gravity case.

Chapter 5

Floating particles

5.1 Governing energy

The case of convex particles in two dimensions was considered in the article [Raphaël, di Meglio, Berger and Calabi (1992)]. Partial results concerning equilibrium criteria were achieved in [McCuan (2007)]. Floating criteria for a long cylinder were given in [Finn (2011)], and in three dimensions and for convex particles in [Finn and Vogel (2009)]. Since we have small bodies in mind where the capillarity plays a larger role than for huge bodies we call our bodies mostly particles. On the other hand capillarity becomes important also for large bodies if the gravity is small.

Whether or not a floating particle or body moves to the boundary of the container wall depends on the contact angles, the geometry and the location of the particle, see [Miersemann (2019)]for the case of a floating beam between two parallel plates.

We will derive equilibrium conditions by forming a two parameter family of admissible comparison configurations. In contrast to the classical problem in capillarity we have to take into account additionally translations and rotations of the particle. Let Ω be the solid particle which can float and which is away from the boundary of the container. Let Ω_s be the fixed solid container. Set $\Sigma = \partial\Omega$ and $\Gamma = \partial\Omega_s$. In this chapter we suppose that the boundaries Σ and Γ are sufficiently smooth. In Chap. 6 we will consider particles with edges.

Let

$$\mathcal{E}(\mathcal{S}, \Sigma) = \sigma|\mathcal{S}| - \sigma\beta_1|\mathcal{W}_1(\mathcal{S})| - \sigma\beta_2|\mathcal{W}_2(\mathcal{S})| \qquad (5.1)$$
$$+ \int_{\Omega_l(\mathcal{S},\Sigma)} F_1(x)\,dx + \int_{\Omega} F_2(x)\,dx$$

be the energy of the problem, where

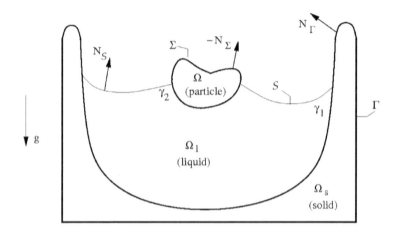

Fig. 5.1   A floating particle, notations

σ surface tension (a positive constant),

$|\mathcal{S}|$ area of the capillary surface $\mathcal{S}$,

$\Omega_l(\mathcal{S}, \Sigma)$ the liquid domain,

β_1 constant (relative) adhesion coefficient between the fluid and the container wall,

β_2 constant (relative) adhesion coefficient between the fluid and the particle,

$\mathcal{W}_1(\mathcal{S})$ wetted part of the container,

$\mathcal{W}_2(\mathcal{S})$ wetted part of the particle,

$F_1 = g\rho_1 x_3$, where ρ_1 is the density of the liquid,

$F_2 = g\rho_2 x_3$, where ρ_2 is the density of the particle.

To simplify the presentation, we suppose that β_1, β_2 and ρ_1, ρ_2 are constants.

For given volume C of the liquid we have additionally the side condition

$$|\Omega_l(\mathcal{S}, \Sigma)| = C, \tag{5.2}$$

where $|\Omega_l(\mathcal{S}, \Sigma)|$ denotes the volume of the liquid domain.

Let $(\mathcal{S}_0, \Sigma_0)$ be a given configuration. Then we will derive necessary conditions such that this configuration defines a minimizer of the associated energy functional subject to a given family of comparison configurations. Assume that $x :\ u \in U \mapsto \mathbb{R}^3$, $u = (u^1, u^2)$ or $u = (\alpha, \beta)$, defines the regular surface $\mathcal{S}_0$, where $U \subset \mathbb{R}^2$ is a double connected domain such that $x(u) \in \Gamma$ if $u \in \partial_1 U$, $x(u) \in \Sigma$ if $u \in \partial_2 U$ and $x(u) \in \mathbb{R}^3 \setminus (\overline{\Omega_s} \cup \overline{\Omega})$ if $u \in U$. Let $y :\ V \mapsto \mathbb{R}^3$, $v = (v^1, v^2)$, be a regular parameter representation of Σ, the boundary of the particle, where $V \subset \mathbb{R}^2$. We choose the parameters such that the normal

$$N_{\mathcal{S}} = \frac{x_{u^1} \wedge x_{u^2}}{|x_{u^1} \wedge x_{u^2}|}$$

is directed *out* of the liquid and that the normal

$$N_{\Sigma} = \frac{y_{v^1} \wedge y_{v^2}}{|y_{v^1} \wedge y_{v^2}|}$$

at $\partial \Omega$ is directed *into* the particle. Moreover, we assume that the normal N_{Γ} at the container wall is directed out of the solid material Ω_s of the container, see Fig. 5.1.

To simplify the presentation we do not consider particles where vapor is enclosed between the capillary surface and the particle.

5.2 Equilibrium conditions

Following [Finn (1986)], Chap. 1, we define for a given configuration $\mathcal{S}_0$, Σ_0, where $\Sigma_0 = \partial \Omega_0$, a one parameter family of admissible comparison surfaces which are in general not yet volume preserving. Set

$$z_0(u, \epsilon) = x(u) + \epsilon \zeta(u) + r(u, \epsilon),$$

$|\epsilon| < \epsilon_0$, the remainder r is continuously differentiable with respect to all arguments, such that $r = O(\epsilon)$ as $\epsilon \to 0$, $z_0(u, \epsilon) \in \mathbb{R}^3 \setminus (\overline{\Omega_s} \cup \overline{\Omega_0})$ if $u \in U$, $z_0(u, \epsilon) \in \Gamma$ if $u \in \partial_1 U$, $z_0(u, \epsilon) \in \Sigma_0$ if $u \in \partial_2 U$, and

$$\zeta(u) = \xi(u) N_{\mathcal{S}_0}(u) + \eta(u) T_{\mathcal{S}_0}(u)$$

is a given vector field. Here $N_{\mathcal{S}_0}$ denotes the unit normal to $\mathcal{S}_0$ pointed to the exterior of the liquid, and $T_{\mathcal{S}_0}$ is a unit tangent field defined on closed strips $U_{i,\delta}$ of $\partial_i U$ of width δ, such that on $\partial \mathcal{S}_0$ the vector $\nu_0 := T_{\mathcal{S}_0}$ is orthogonal to $\partial_1 \mathcal{S}_0$, $\partial_2 \mathcal{S}_0$, resp., and points to the exterior of the liquid, see Fig. 5.2. We assume that ξ and η are sufficiently regular on $\overline{U}$, supp $\eta \subset U_{1,\delta} \cup U_{2,\delta}$, and $\xi^2 + \eta^2 \le 1$. Define on $\partial_1 \mathcal{S}_0 = \{x(u) : u \in \partial_1 U\}$ the angle $\gamma_1 \in [0, \pi]$ by $\cos \gamma_1 = N_{\Gamma} \cdot N_{\mathcal{S}_0}$ and on $\partial_2 \mathcal{S}_0 = \{x(u) : u \in \partial_2 U\}$ the angle $\gamma_2 \in [0, \pi]$ by $\cos \gamma_2 = -N_{\Sigma_0} \cdot N_{\mathcal{S}_0}$. We recall that N_{Σ_0} is directed *into* the particle. These angles depend on $P \in \partial \mathcal{S}_0$. Later on we will see that these angles are constants. Since $\langle \zeta, N_{\Gamma} \rangle = 0$ at $\partial \mathcal{S}_0$ it follows

$$\xi(u) \cos \gamma_1 - \eta(u) \sin \gamma_1 = 0, \quad u \in \partial_1 U, \tag{5.3}$$

$$\xi(u) \cos \gamma_2 - \eta(u) \sin \gamma_2 = 0, \quad u \in \partial_2 U. \tag{5.4}$$

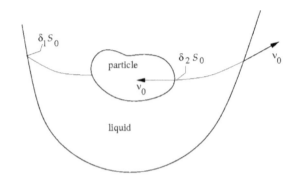

Fig. 5.2   Normals ν_0, notations

Thus for unchanged Σ_0 the family $z_0(u, \epsilon)$, where ζ satisfies equations (5.3) and (5.4), defines a family of admissible comparison configurations. Such a family exists provided the boundaries of the container and of the particle are sufficiently regular. The proof exploits an implicit function theorem, we omit the details.

To incorporate rigid motions of the particle we form a family of admissible comparison configurations, which are in general not yet volume preserving, as follows. Let $R(\omega) = R_1(\omega_1)R_2(\omega_3)R_3(\omega_3)$ be a rotation matrix, where $R_i(\omega_i)$ defines a rotation around the x_i-axis with angle ω_i. These rotations are given by

$$R_1(\tau) = \begin{pmatrix} 1 & 0 & 0 \\ 0 & \cos\tau & -\sin\tau \\ 0 & \sin\tau & \cos\tau \end{pmatrix},$$

$$R_2(\tau) = \begin{pmatrix} \cos\tau & 0 & \sin\tau \\ 0 & 1 & 0 \\ -\sin\tau & 0 & \cos\tau \end{pmatrix},$$

$$R_3(\tau) = \begin{pmatrix} \cos\tau & -\sin\tau & 0 \\ \sin\tau & \cos\tau & 0 \\ 0 & 0 & 1 \end{pmatrix}.$$

Let $\chi(u)$, $u \in \overline{U}$, be a sufficiently regular function such that

$$\chi(u) = \begin{cases} 0 & : \quad u \in \partial_1 U \\ 1 & : \quad u \in \partial_2 U \end{cases}.$$

Set $V_0 = \{v \in V : y(v) \in \mathcal{W}_2(\mathcal{S}_0)\}$, where $\mathcal{W}_2(\mathcal{S}_0)$ is the wetted part of the unperturbed particle. For given ζ and α, $a \in \mathbb{R}^3$ we define a two parameter family of admissible configurations, not yet volume preserving in general, $(\mathcal{S}(\mu, \epsilon), \Sigma(\mu, \epsilon))$, where the capillary surface $\mathcal{S}(\mu, \epsilon)$ is given by

$$z(u, \mu, \epsilon) = (1 - \chi(u))z_0(u, \epsilon) + \chi(u)\left(R(\mu\alpha)z_0(u, \epsilon) + \mu a\right),$$

$u \in U$, and the wetted part $\Sigma(\mu, \epsilon)$ of the moved particle is defined by

$$z_\Sigma(v, \mu) = R(\mu\alpha)y(v) + \mu a, \quad v \in V(\epsilon),$$

where

$$V(\epsilon) = \{v \in V : y(v) \in \mathcal{W}_2(\mathcal{S}^*(0, \epsilon))\}$$

is the parameter domain of Σ which defines the wetted part of the unchanged particle Ω_0 subject to $\mathcal{S}(0, \epsilon)$. The moved particle is defined by

$$\Omega(\mu) = \{y \in \mathbb{R}^3 : y = R(\mu\alpha)x + \mu a, \ x \in \Omega_0\}.$$

To get a volume preserving admissible family of configurations we replace $z(u, \mu, \epsilon)$ through

$$z^*(u, \mu, \epsilon, q) := z(u, \mu, \epsilon) + q\xi_0(u)N_0(u),$$

where $q \in \mathbb{R}$, $N_0 = N_{\mathcal{S}_0}$, supp $\xi_0 \subset U$, supp $\xi_0 \cap$ supp $\chi = \emptyset$ and

$$\int_{\mathcal{S}_0} \xi_0(u) \, dA = 1.$$

Then $(\mathcal{S}^*(\mu, \epsilon, q), \Sigma(\mu, \epsilon))$, where $\mathcal{S}^*(\mu, \epsilon, q)$ is defined by $z^*(u, \mu, \epsilon, q)$, provides a family of admissible configurations, in general not yet volume preserving. Set

$$X = \{(\zeta, \alpha, a) \in C^1(U) \cap C^0(\overline{U}) \times \mathbb{R}^3 \times \mathbb{R}^3 : \zeta \text{ satisfies (5.5) and (5.4)}\}$$

and let X_0 be the subspace where $w \in X_0$ satisfies the side conditions

$$\int_{\mathcal{S}_0} \xi \, dA = 0 \tag{5.5}$$

and

$$\int_{\mathcal{S}_0} \chi(u)\langle N_{\mathcal{S}_0}(u), (\alpha_i R_{\alpha_i}(0))x(u) + a \rangle \, dA \tag{5.6}$$

$$+ \int_{\mathcal{W}_2(\mathcal{S}_0)} \langle N_{\Sigma_0}(v), (\alpha_i R_{\alpha_i}(0))y(v) + a \rangle \, dA = 0.$$

From Lemma 5.1 of the appendix to this chapter we have that for given $w \in X_0$ there exists a regular function $q = q(\mu, \epsilon)$ such that the family of admissible configurations given by $(\mathcal{S}^*(\mu, \epsilon, q(\mu, \epsilon)), \Sigma(\mu, \epsilon))$, is volume preserving for all $(\mu, \epsilon) \in \mathbb{R}^2$, $\mu^2 + \epsilon^2 < c_0^2$. The function $q(\mu, \epsilon)$ satisfies $q(0, 0) = 0$, $q_\mu(0, 0) = 0$ and $q_\epsilon(0, 0) = 0$.

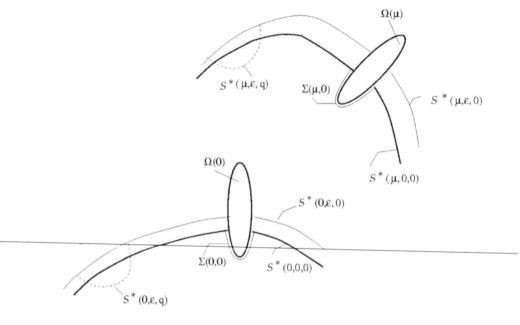

Fig. 5.3   Volume preserving motion of the particle, $q = q(\mu, \epsilon)$.

In the following we assume for ζ and the remainder $r(u, \epsilon)$ in the definition of $z_0(u, \epsilon)$ that all integrals below exist. Set $c = (\mu, \epsilon)$ and $S^*(c) := S^*(\mu, \epsilon, q(\mu, \epsilon))$. Suppose that for fixed volume C of the liquid the above family $(S^*(c), \Sigma(c))$ of volume preserving admissible configurations satisfies for all c, $|c| < c_0$,

$$\mathcal{E}(S^*(c), \Sigma(c)) \geq \mathcal{E}(S^*(0), \Sigma(0)).$$

Consequently,

$$[\nabla_c \mathcal{E}(S^*(c), \Sigma(c))]_{c=0} = 0, \qquad (5.7)$$

$$[\nabla_c |\Omega_l(S^*(c), \Sigma(c))|]_{c=0} = 0. \qquad (5.8)$$

Equation (5.8) defines the subspace X_0. Set

$$\mathcal{L}(S, \Sigma, \lambda) = \mathcal{E}(S, \Sigma) + \lambda(|\Omega_l(S, \Sigma)| - C).$$

From a Lagrange multiplier rule, see Chap. 10, it follows that there exists a real λ_0 such that the family $(S(c), \Sigma(c))$, in general not volume preserving, where $S(c)$ is given by $z(u, \mu, \epsilon)$, satisfies

$$[\nabla_c \mathcal{L}(S(c), \Sigma(c), \lambda_0)]_{c=0} = 0.$$

We recall that

$$\mathcal{L}(\mathcal{S}(c), \Sigma(c), \lambda_0) = \sigma|\mathcal{S}(c)| - \sigma\beta_1|\mathcal{W}_1(\mathcal{S}(c))| - \sigma\beta_2|\mathcal{W}_2(\mathcal{S}(c))|$$
$$+ \int_{\Omega_l(\mathcal{S}(c),\Sigma(c))} F_1(x)\, dx + \int_{\Omega(\mu)} F_2(x)\, dx$$
$$+ \lambda_0 \left(|\Omega_l(\mathcal{S}(c), \Sigma(c))| - C\right). \tag{5.9}$$

The argument c indicates that the function or vector in consideration is related to the perturbed surface $\mathcal{S}(c)$. Set

$$W(c) = \sqrt{E(c)G(c) - F^2(c)},$$

where E, F, G are the coefficients of the first fundamental form of $\mathcal{S}(c)$ defined by $z(u, c) := z(u, \mu, \epsilon)$.

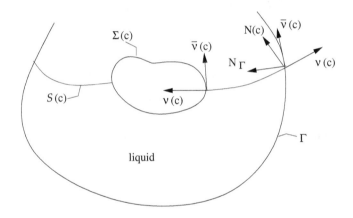

Fig. 5.4   Normals in consideration

By $\nu(u, c)$, resp. $\bar{\nu}(u, c)$, we denote the exterior normal to $\partial\mathcal{S}(c)$ in $\mathcal{S}(c)$, resp., at the container wall Γ or at the boundary $\partial\Omega(\mu)$ of the particle at $P \in \partial\mathcal{S}(c)$, see Fig. 5.4. If $P \in \partial\mathcal{S}(c)$, then the vectors N_Γ, $N(u, c)$, $\nu(u, c)$ and $\bar{\nu}(u, c)$ are all in the same plane since $\partial\mathcal{S}(c)$ is a curve on $\mathcal{S}(c)$ as well as on the container wall Γ or on $\partial\Omega(\mu)$. Here and in the following we set $N(u, c) = N_{\mathcal{S}(c)}$ at $u \in U$.

Let $H_0 = H(u, 0)$, $\nu_0 = \nu(u, 0)$, $\bar{\nu}_0 = \bar{\nu}(\tau, 0)$, where $u \in \partial U$, $N_0 = N(u, 0)$, which is equal $N_{\mathcal{S}_0}$ at $u \in U$, and $N_{\Sigma_0} = N_\Sigma(v, 0)$. We recall that $\mathcal{S}_0 = \mathcal{S}(0)$, where $\mathcal{S}_0$, Σ_0 is the given configuration under consideration.

Equation

$$\mathcal{L}_\epsilon(\mathcal{S}_0, \Sigma_0, \lambda_0) := [\mathcal{L}_\epsilon(\mathcal{S}(c), \Sigma(c), \lambda_0)]_{c=0} = 0$$

implies the well known equilibrium conditions of a liquid in a fixed container, see [Finn (1986)], p. 10. We obtain from the corollaries to Lemmas 5.1–5.5 the formula

$$[\mathcal{L}_\epsilon(\mathcal{S}(c),\Sigma(c),\lambda_0)]_{c=0} = -2\sigma \int_{\mathcal{S}_0} H_0\xi\, dA + \sigma \int_{\partial_1\mathcal{S}_0} \langle\zeta,\nu_0 - \beta_1\bar{\nu}_0\rangle\, ds$$
$$+ \sigma \int_{\partial_2\mathcal{S}_0} \langle\zeta,\nu_0 - \beta_2\bar{\nu}_0\rangle\, ds + \int_{\mathcal{S}_0} F_1(x(u))\xi\, dA$$
$$+ \lambda_0 \int_{\mathcal{S}_0} \xi\, dA.$$

Equilibrium conditions follow from $\mathcal{L}_\epsilon(\mathcal{S}_0,\Sigma_0,\lambda_0) = 0$ for all $\zeta = \xi N_0 + \eta T_{\mathcal{S}_0}$, where $T_{\mathcal{S}_0} = \nu_0$ on $\partial\mathcal{S}_0$, and $\xi\cos\gamma_k - \eta\sin\gamma_k = 0$ on $\partial_k\mathcal{S}_0$. Consider a subset of these ζ's where $\zeta = 0$ on $\partial\mathcal{S}_0$, then we get the equation

$$-2\sigma H_0 + F_1 + \lambda_0 = 0 \tag{5.10}$$

on $\mathcal{S}_0$. Consequently,

$$\int_{\partial_k\mathcal{S}_0} \langle\zeta,\nu_0 - \beta_k\bar{\nu}_0\rangle\, ds = 0$$

for all ζ defined above. Thus we get the boundary conditions

$$\cos\gamma_k = \beta_k.$$

on $\partial_k\mathcal{S}_0$ since we have on $\partial_k\mathcal{S}_0$

$$\langle\zeta,\nu_0 - \beta_k\bar{\nu}_0\rangle = \xi\langle N_0,\nu_0 - \beta_k\bar{\nu}_0\rangle + \eta\langle\nu_0,\nu_0 - \beta_k\bar{\nu}_0\rangle$$
$$= -\beta_k\xi\langle N_0,\bar{\nu}_0\rangle + \eta(1 - \beta_k\langle\nu,\bar{\nu}_0\rangle)$$
$$= -\xi\beta_k\sin\gamma_k + \eta(1 - \beta_k\cos\gamma_k).$$

Following [Finn (1986)], p. 10, we set $\xi = \tau_k\sin\gamma_k$ and $\eta = \tau_k\cos\gamma_k$, where $\tau_k \in C(\partial_k U)$. Then $\xi\cos\gamma_k - \eta\sin\gamma_k = 0$ is satisfied for all τ_k. Then

$$\langle\zeta,\nu_0 - \beta_k\bar{\nu}_0\rangle = \tau_k(\cos\gamma_k - \beta_k). \tag{5.11}$$

In contrast to the classical capillary problem with a fixed container we get additional conditions from the fact that the particle can move in the liquid. These conditions are given by equation

$$\mathcal{L}_\mu(\mathcal{S}_0,\Sigma_0,\lambda_0) := [\mathcal{L}_\mu(\mathcal{S}(c),\Sigma(c),\lambda_0)]_{c=0} = 0.$$

We have, see the corollaries to Lemmas 5.1–5.5,

$$[\mathcal{L}_\mu(\mathcal{S}(c), \Sigma(c), \lambda_0)]_{c=0} = -2\sigma \int_{\mathcal{S}_0} H_0(u)\langle N_0(u), z_\mu(u,0,0)\rangle \, dA$$

$$+ \sigma \int_{\partial_1 \mathcal{S}_0} \langle \nu_0(u), z_\mu(u,0,0)\rangle \, ds + \sigma \int_{\partial_2 \mathcal{S}_0} \langle \nu_0(u), z_\mu(u,0,0)\rangle \, ds$$

$$+ \int_{\mathcal{S}_0} F_1(x(u))\langle N_0(u), z_\mu(u,0,0)\rangle \, dA$$

$$+ \int_{\mathcal{W}_2(\mathcal{S}_0)} F_1(y(v))\langle N_{\Sigma_0}(v), z_{\Sigma,\mu}(v,0)\rangle \, dA$$

$$+ \int_{\Omega_0} \langle \nabla F_2(y), (\alpha_i R_{\alpha_i}(0))y + a\rangle \, dy$$

$$+ \lambda_0 \left(\int_{\mathcal{S}_0} \langle N_0(u), z_\mu(u,0,0)\rangle \, dA + \int_{\mathcal{W}_2(\mathcal{S}_0)} \langle N_{\Sigma_0}(v), z_{\Sigma,\mu}(v,0)\rangle \, dA \right)$$

Since

$$z_{\Sigma,\mu}(v,0) = (\alpha_i R_{\alpha_i}(0))y(v) + a,$$

$$z_\mu(u,0,0) = \chi(u)\left((\alpha_i R_{\alpha_i}(0))y(v) + a\right),$$

where $\chi(u) = 1$ on $\partial_2 U$ and $\chi(u) = 0$ on $\partial_1 U$, we find from $\mathcal{L}_\mu(\mathcal{S}_0, \Sigma_0, \lambda_0) = 0$ and, using equation (5.10), that

$$\sigma \int_{\partial_2 \mathcal{S}_0} \langle \nu_0(u), (\alpha_i R_{\alpha_i}(0))x(u) + a\rangle \, ds$$

$$+ \int_{\mathcal{W}_2(\mathcal{S}_0)} F_1(y(v))\langle N_{\Sigma_0}(v), (\alpha_i R_{\alpha_i}(0))y(v) + a\rangle \, dA$$

$$+ \int_{\Omega_0} \langle \nabla F_2(y), (\alpha_i R_{\alpha_i}(0))y + a\rangle \, dy$$

$$+ \lambda_0 \int_{\mathcal{W}_2(\mathcal{S}_0)} \langle N_{\Sigma_0}(v), (\alpha_i R_{\alpha_i}(0))y(v) + a\rangle \, dA = 0 \qquad (5.12)$$

for all α, $a \in \mathbb{R}^3$.

Let $\mathcal{S}(c)$, $\Sigma(c)$ be the family of volume preserving admissible configurations defined above. Summarizing the previous formulas, we have shown

Theorem 5.1. *Suppose that the configuration $\mathcal{S}_0$, Σ_0 defines a local minimum of $\mathcal{E}(\mathcal{S}(c)), \Sigma(c))$ under the side condition $|\Omega_l(\mathcal{S}(c), \Sigma(c))| = C$, $C > 0$ is a given constant. Then there exists a real constant λ_0 such that*

$$-2\sigma H_0 + F_1 + \lambda_0 = 0 \quad on\ \mathcal{S}_0, \qquad (5.13)$$

$$\cos\gamma_1 = \beta_1 \quad at\ \partial_1 \mathcal{S}_0, \qquad (5.14)$$

$$\cos\gamma_2 = \beta_2 \quad at\ \partial_2 \mathcal{S}_0, \qquad (5.15)$$

$$\sigma \int_{\partial_2 S_0} \nu_0(u)\, ds + \int_{\mathcal{W}_2(S_0)} F_1(y(v)) N_{\Sigma_0}(v)\, dA \tag{5.16}$$

$$+ \int_{\Omega_0} \nabla F_2(y)\, dy + \lambda_0 \int_{\mathcal{W}_2(S_0)} N_{\Sigma_0}(v)\, dA = 0,$$

$$\sigma \int_{\partial_2 S_0} \langle R_{\alpha_i}(0) x(u), \nu_0(u) \rangle\, ds \tag{5.17}$$

$$+ \int_{\mathcal{W}_2(S_0)} F_1(y(v)) \langle R_{\alpha_i}(0) y(v), N_{\Sigma_0}(v) \rangle\, dA + \int_{\Omega_0} \langle \nabla F_2(y), R_{\alpha_i}(0) y \rangle\, dy$$

$$+ \lambda_0 \int_{\mathcal{W}_2(S_0)} \langle R_{\alpha_i}(0) y(v), N_{\Sigma_0}(v) \rangle\, dA = 0,$$

where $i = 1,\ 2,\ 3$.

Remark 5.1. In general, if merely equations (5.13)–(5.15) are satisfied, then the force $\mathcal{F} = (\mathcal{F}_1, \mathcal{F}_2, \mathcal{F}_3)$ and the torques (moments of force) $\mathcal{M}_k$ with respect to the x_k-axis keep the body in an equilibrium if the following equations are satisfied:

$$\mathcal{F}_k = \sigma \int_{\partial_2 S_0} \langle \nu_0(u), e_k \rangle\, ds + \int_{\mathcal{W}_2(S_0)} F_1(y(v)) \langle N_{\Sigma_0}(v), e_k \rangle\, dA$$

$$+ \int_{\Omega_0} \langle \nabla F_2(y), e_k \rangle\, dy + \lambda_0 \int_{\mathcal{W}_2(S_0)} \langle N_{\Sigma_0}(v), e_k \rangle\, dA.$$

Here we denote by e_k the unit vectors of the standard basis in $\mathbb{R}^3$, and

$$\mathcal{M}_k = \sigma \int_{\partial_2 S_0} \langle R_{\alpha_k}(0) x(u), \nu_0(u) \rangle\, ds$$

$$+ \int_{\mathcal{W}_2(S_0)} F_1(y(v)) \langle R_{\alpha_k}(0) y(v), N_{\Sigma_0}(v) \rangle\, dA$$

$$+ \int_{\Omega_0} \langle \nabla F_2(y), R_{\alpha_k}(0) y \rangle\, dy + \lambda_0 \int_{\mathcal{W}_2(S_0)} \langle R_{\alpha_k}(0) y(v), N_{\Sigma_0}(v) \rangle\, dA.$$

5.2.1 Restricted movements

Now we consider some cases of restricted motions of the body.

5.2.1.1 Vertical movement

Suppose that the particle can move in the x_3-direction only and no rotation is allowed, see Fig. 5.5 Then it follows from the above remark that the vertical force $\mathcal{F}_3$ directed upwards keeps the particle in an equilibrium provided that equations (5.13)–(5.15) and

$$
\mathcal{F}_3 = \sigma \int_{\partial_2 S_0} \langle \nu_0(u), e_3 \rangle \, ds + \int_{\mathcal{W}_2(S_0)} F_1(y(v)) \langle N_{\Sigma_0}(v), e_3 \rangle \, dA
$$
$$
+ \int_{\Omega_0} \langle \nabla F_2(y), e_3 \rangle \, dy + \lambda_0 \int_{\mathcal{W}_2(S_0)} \langle N_{\Sigma_0}(v), e_3 \rangle \, dA
$$

are satisfied.

If the body can rotate around the x_3-axis additionally, then the torque $\mathcal{M}_3$ vanishes, see the above remark where the torques are defined.

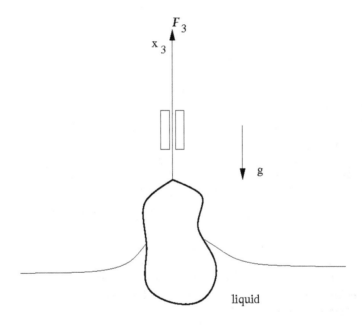

Fig. 5.5   Vertical movement

Remark 5.2. The buoyancy $\mathcal{B}$ of the body is given by

$$
\mathcal{B} = - \int_{\mathcal{W}_2(S_0)} F_1(y(v)) \langle N_{\Sigma_0}(v), e_3 \rangle \, dA.
$$

We recall that N_{Σ_0} denotes the interior normal at the boundary of the body.

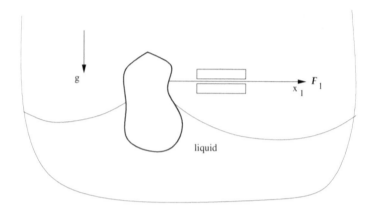

Fig. 5.6   Horizontal movement

5.2.1.2   *Horizontal movements*

Assume that the body can move in the x_1-direction only and no rotation is allowed, see Fig. 5.6.

Then the horizontal force $\mathcal{F}_1$ directed in the positive x_1-axis keeps the body in an equilibrium if equations (5.13)–(5.15) and

$$\mathcal{F}_1 = \sigma \int_{\partial_2 S_0} \langle \nu_0(u), e_1 \rangle \, ds + g\rho_1 \int_{\mathcal{W}_2(S_0)} y_3(v) \langle N_{\Sigma_0}(v), e_1 \rangle \, dA$$

$$+ \lambda_0 \int_{\mathcal{W}_2(S_0)} \langle N_{\Sigma_0}(v), e_1 \rangle \, dA$$

are satisfied, where $y(v) = (y_1(v), y_2(v), y_3(v))$, $v \in V$, defines the boundary of the particle. We recall that ρ_1 denotes the density of the liquid.

It was firstly shown in [Finn (2010)], see also [Finn (2013); Aspley, He and McCuan (2015)], that parallel plates which are parallel to the gravity force can attract or repel each other depending on the distance of the plates and on the two contact angles of the faces between the plates. Here we consider a plate parallel to a fixed vertical wall, and we assume that the plate can move in the x_1-direction only, see Fig. 5.7. Since there is no volume constraint in this case, we have $\lambda_0 = 0$. Then, see the above formulas, the force $\mathcal{F}_1$ is given by

$$\mathcal{F}_1 = \sigma \int_{\partial_2 S_0} \langle \nu_0(v), e_1 \rangle \, ds + g\rho_1 \int_{\mathcal{W}_2(S_0)} y_3(v) \langle N_{\Sigma_0}(v), e_1 \rangle \, dA, \quad (5.18)$$

where N_{Σ_0} is the interior normal, if it exists, at the boundary of the plate, and $\mathcal{F}_1$ is positive if the wall attracts the plate and negative if the wall

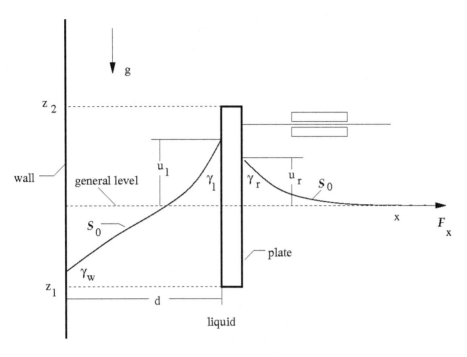

Fig. 5.7 Floating plate

repels the plate. Consider the case that the plate is infinitely long in both y-directions, and denote by $\mathcal{F}_1^L$ the force acting on a part of the plate of length L in the y-direction. Then we get from formula (5.18) that

$$\mathcal{F}_1^L = L\sigma(\sin\gamma_l - \sin\gamma_r) + \frac{1}{2}g\rho_1 L(u_l^2 - u_r^2),$$

where u_l is the ascent of the liquid at the left part of the plate and u_r is the ascent at the right plane, both measured from the general level of the liquid which is the line given by $z = 0$. We assume that the plate is sufficiently height and deep such that the inequalities $z_1 < u_l$, $u_r < z_2$ are satisfied. Since, see Sec. 3.3.1,

$$u_r^2 = \frac{2\sigma}{g\rho_1}(1 - \sin\gamma_r),$$

we get

$$\mathcal{F}_1^L = L\left(\sigma(\sin\gamma_l - 1) + \frac{g\rho_1}{2}u_l^2\right),$$

where $u_l = u_l(d, \gamma_w, \gamma_l)$. Assume that $0 \leq \gamma_l \leq \pi$ and let h be the ascent of the liquid at a vertical wall with contact angle γ_l. Then the plate will be repelled if $|u_l| < h$, and attracted if $|u_l| > h$.

Let $\gamma_w = \gamma_l =: \gamma$, then, see an exercise of Chap. 7,

$$u_1 = \frac{2}{\kappa d} \cos \gamma + O(d)$$

as $d \to 0$. Then the attracting force tends to infinity if d tends to zero, provided that $\gamma \neq 0$.

Let ψ be the inclination angle of the curve at $(x, u(x))$ and set

$$C = \cos \psi + \frac{\kappa}{2} u^2. \tag{5.19}$$

If $u(x)$ is a solution of the one-dimensional capillary equation (3.11) on the interval $[0, d]$, then C is an integral, i. e., C is constant on $[0, d]$, see an exercise. Consequently,

$$\mathcal{F}_1^L = L\sigma(C - 1).$$

We recall that $\kappa = g\rho_1/\sigma$. From formula (5.19) we get that the plate will be attracted if $0 \leq \gamma_w \leq \pi/2$ and $0 \leq \gamma_1 \leq \pi/2$, since there is a zero of $u'(x)$ in the interval $[0, d]$, and the plate will be repelled if $\gamma_w = \pi - \gamma_l$, $0 \leq \gamma_l \leq \pi$, since there is a zero of $u(x)$ in the interval $[0, d]$.

In general, for other angles and distances the behavior is more complicated, see [Finn (2010, 2006); Aspley, He and McCuan (2015)].

Since C is an integral, we get

$$C = \sin \gamma_w + \frac{\kappa}{2} u_w^2 = \sin \gamma_l + \frac{\kappa}{2} u_l^2,$$

where u_w is the ascent of the liquid at the wall and u_l is the ascent at the plate. Let $v(B, \gamma_w, \gamma_l, x)$ be the solution of the one-dimensional capillary boundary value problem on the interval $[0, 1]$

$$\left(\frac{v'(x)}{\sqrt{1 + v'(x)^2}} \right)' = Bv(x),$$

$$\lim_{x \to +0} \frac{v'(x)}{\sqrt{1 + v'(x)^2}} = -\cos \gamma_w,$$

$$\lim_{x \to 1-0} \frac{v'(x)}{\sqrt{1 + v'(x)^2}} = \cos \gamma_l.$$

Then

$$u_w = d\, v(\kappa d^2, \gamma_w, \gamma_l, 0),$$
$$u_l = d\, v(\kappa d^2, \gamma_w, \gamma_l, 1).$$

One obtains these quantities by using a numerical method prescribed in Sec. 7.2.8. Consequently, one can find out for given data d, γ_w, γ_l whether or not the plate is repelled or attracted from the wall, and which force is acting on the plate.

Example 5.1. Let $\sigma = 73.8\ mN/m$, which is approximately the surface tension of water with a temperature of about 20° Celsius. Then one gets approximately $\kappa = 13.3\ cm^{-2}$. Assume the distance of the plates is $0.1\ cm$. Suppose that $\gamma_l = 1^0$ and $\gamma_w = 179^0$, then $\mathcal{F}_1 = -50.9\ mN$, and if $\gamma_l = 1^0$ and $\gamma_w = 1^0$, then $\mathcal{F}_1 = 959.3\ mN$.

Remark 5.3. Consider the case, see [Finn (2010, 2013); Aspley, He and McCuan (2015)], that the left plate can flow too and let $\mathcal{F}$ be the net force, i. e., the difference between the forces acting on the right and on the left plate. From formula (5.18) it follows that

$$\mathcal{F} = 2L\sigma(C - 1),$$

i. e., the attracting/repelling force is twice of the related force if the left plate is a fixed vertical wall.

The problem of a floating plate or beam between two fixed parallel plates is discussed in [Miersemann (2019)]. Probably, this problem provides hints to the behavior of a floating ball in a container with circular cross section. Kitchen sink experiments show that a ping-pong ball does not stay in the middle of the tube in contrast to the case where the capillary surface hangs on the interior edge of the tube and makes a contact angle larger than $\pi/2$, see Fig. 1.12.

Instead of a floating ball we consider a floating beam between two fixed parallel plates, see Fig. 5.8. It is assumed that the beam is indefinitely long and is floating parallel to the plates.

Moreover, it is supposed that the beam can move perpendicular to the plates only, maybe along a bar, see Fig. 5.8. The acting force $\mathcal{F}_x^L$ in x-direction and per length unit L is defined through the equilibrium condition

$$\sigma \int_{\partial S} \langle \nu, e_1 \rangle \, ds + g\rho_1 \int_{\mathcal{W}(S)} y_3(v)\langle N_\Sigma, e_1 \rangle \, dA + g\rho_2 |\Omega| = \mathcal{F}_x^L.$$

Consequently,

$$\mathcal{F}_x^L = L\sigma(\sin \gamma_l - \sin \gamma_r) + \frac{1}{2}g\rho_1 L(u_l^2 - u_r^2),$$

and if $\gamma_l = \gamma_r =: \gamma_p$, then we get

$$\mathcal{F}_x^L = \frac{1}{2}g\rho_1 L(u_l^2 - u_r^2). \tag{5.20}$$

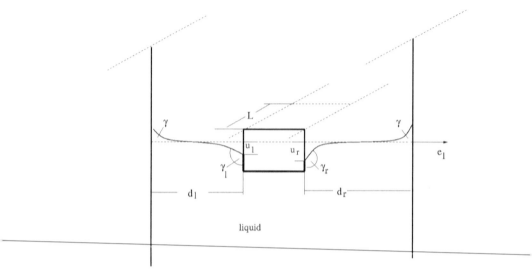

Fig. 5.8 Floating beam

Assume that the boundary contact angles satisfy $0 \leq \gamma_p < 90^\circ$ and $0 \leq \gamma < 90^\circ$. Then, provided that $d_r < d_l$, we get the inequality $\mathcal{F}_x^L < 0$ since $0 < u_l < u_r$. This inequality is a consequence of a comparison principle for unbounded domains, see [Finn and Hwang (1989)]. This means that the beam moves to the right since we must push the beam to the left in order to keep it in an equilibrium.

The same assertion holds in the case if $90^\circ < \gamma_p \leq 180^\circ$ and $90^\circ < \gamma \leq 180^\circ$. Since, if $d_r < d_l$, one finds from the comparison principle again $u_r < u_l < 0$.

The behavior is more complicated in the case of other contact angles, see [Finn (2010)] for the case of parallel plates. One finds the ascents u_l and u_r from the numerical solution, see Chap. 7, of the following nonlinear boundary problems (5.21)–(5.23)

$$\left(\frac{u'(r)}{\sqrt{1 + (u'(r))^2}} \right)' = \kappa\, u(r), \quad 0 < r < d, \tag{5.21}$$

$$\frac{u'(r)}{\sqrt{1 + (u'(r))^2}} = -\cos\gamma_p, \quad r = 0, \tag{5.22}$$

$$\frac{u'(r)}{\sqrt{1 + (u'(r))^2}} = \cos\gamma, \quad r = d. \tag{5.23}$$

Example 5.2. Let $\kappa := 13.3\,cm^{-2}$, $g := 9.81\,m/s^2$ and $\rho_1 := 10^{-3}/cm^3$: If $d_l = 1.8\,cm$ $d_r = 0.2\,cm$, $\gamma_p = 40°$ and $\gamma = 178°$, $L = 1\,cm$. Then $\mathcal{F}_x^L = 0.216\,mN$, and if $d_l = 1.1\,cm$ $d_r = 0.9\,cm$ then $\mathcal{F}_x^L = 0.013\,mN$. That is, in both cases the beam flows into the middle. If d_l, d_r and γ_p as above, but $\gamma = 100°$, then one finds in the first case from above that $\mathcal{F}_x^L = -0.156\,mN$, that means the beam moves to the right. In the second case the beam flows to the middle since $\mathcal{F}_x^L = 0.0027\,mN$.

5.2.1.3 *Rotation around the x_3-axis*

Suppose that the particle can rotate around the x_3-axis only and no translation is allowed, see Fig. 5.9. Then the particle is in an equilibrium if equations (5.13)–(5.15) are satisfied and if the torque $\mathcal{M}_3$ vanishes, i. e.,

$$\sigma \int_{\partial_2 S_0} \langle R_{\alpha_3}(0)x(u), \nu_0(u) \rangle \, ds + g\rho_1 \int_{\mathcal{W}_2(S_0)} y_3(v) \langle R_{\alpha_3}(0)y(v), N_{\Sigma_0}(v) \rangle \, dA$$

$$+ \lambda_0 \int_{\mathcal{W}_2(S_0)} \langle R_{\alpha_3}(0)y(v), N_{\Sigma_0}(v) \rangle \, dA = 0,$$

where

$$R_{\alpha_3}(0) = \begin{pmatrix} 0 & -1 & 0 \\ 1 & 0 & 0 \\ 0 & 0 & 0 \end{pmatrix}.$$

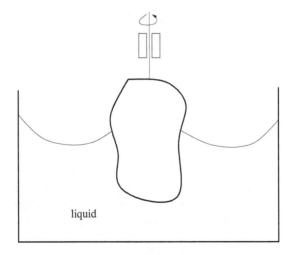

liquid

Fig. 5.9 Rotation around the x_3-axis

Remark 5.4. If the volume of liquid is infinite, then $\lambda_0 = 0$ in the above formula.

5.2.2 *Explicit solutions*

Here we consider examples which lead to explicit results.

5.2.2.1 *Floating ball, horizontal capillary surfaces*

Consider a floating ball in a cylindrical tube, not necessarily with circular cross section. Assume that the liquid makes a contact angle of $\pi/2$ with the container wall. Let ρ_1 be the constant density of the liquid and ρ_2 the constant density of the ball. We suppose that $0 < \rho_2 < \rho_1$. In general, for given $\gamma_2 \in (0, \pi)$ the capillary surface is not a subset of a plane. On the other hand, we have

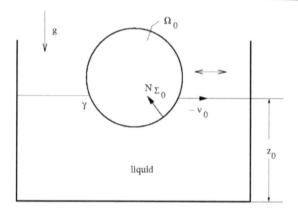

Fig. 5.10   Horizontal capillary surface

Proposition 5.1. *Suppose that the gravity g is positive and directed downwards in the direction of the negative x_3-axis, see Fig. 5.10. Then there exists a unique contact angle γ_2 such that the horizontal capillary surface S_0 satisfies the equilibrium criteria of the theorem. Moreover, we have $0 < \gamma_2 < \pi/2$ if $2\rho_2 > \rho_1$, $\gamma_2 = \pi/2$ if $2\rho_2 = \rho_1$ and $\pi/2 < \gamma_2 < \pi$ if $2\rho_2 < \rho_1$. Every ball moved horizontally and stays away from the boundary of the container defines an equilibrium.*

Proof. Suppose that S_0 is a subset of the plane $x_3 = z_0$, see Fig. 5.10 for notations, and makes a contact angle γ with the ball. From the equilibrium condition (5.13) we get $\lambda_0 = -g_1 \rho_1 z_0$. Since $\int_{\partial_2 S_0} \nu_0 \, ds = 0$, the equilibrium condition (5.16) reduces to

$$g\rho_1 \int_{\mathcal{W}_2(S_0)} y_3(v) N_{\Sigma_0}(v) \, dA + g\rho_2 \int_{\Omega_0} \begin{pmatrix} 0 \\ 0 \\ 1 \end{pmatrix} dx + \lambda_0 \int_{\mathcal{W}_2(S_0)} N_{\Sigma_0}(v) \, dA = 0.$$

This equation is satisfied if γ is a zero of $f(\gamma)$ defined by

$$f(\gamma) = \rho_1 \left(\frac{1}{3} \cos^3 \gamma - \cos \gamma - \frac{2}{3} \right) + \frac{4}{3} \rho_2.$$

The assertions of the proposition follow since $f(0) = -4/3 \, (\rho_1 - \rho_2)$, $f(\pi) = 4/3 \, \rho_2$ and

$$f(\pi/2) \begin{cases} = 0 & , & 2\rho_2 = \rho_1 \\ > 0 & , & 2\rho_2 > 2\rho_1 \\ < 0 & , & 2\rho_2 < 2\rho_1 \end{cases}$$

Since $f'(\gamma) = \rho_1 \sin^3 \gamma$, we get uniqueness of the contact angle. The equilibrium criterion (5.17) follows from the symmetry properties of the particle or by formal calculations. In fact, in the case of a ball as a particle the equilibrium condition (5.17) is superfluous because we can set $\alpha = 0$ in the definition of the admissible configurations. □

5.2.2.2 *Zero gravity*

Consider a floating ball in the zero gravity case. Then for *every* given contact angle $\gamma_2 \in (0, \pi)$ there is a horizontal capillary surface since the equilibrium condition (5.13) shows that $\lambda_0 = 0$, and condition (5.17) is satisfied. This behavior for floating balls under zero gravity was firstly discovered by [Finn (2006)]. An important consequence is a further discovery of Finn that Young's surface tension diagram fails for floating particles, see [Finn, McCuan and Wente (2012)].

Suppose that the particle contains a rotationally symmetric part and that all angles between this part and planes which intersect the axis of rotation perpendicular are different from each other. Assume the contact angle between the particle and the liquid is in this set of angles. Then there is a horizontal surface in equilibrium, see Fig. 5.11. One expects that one can keep such a particle vertically in zero gravity provided that a tilted particle makes a non constant contact angle with planes perpendicular to the axis of symmetry. For example, this assumption is satisfied for a rotationally ellipsoid different from a ball.

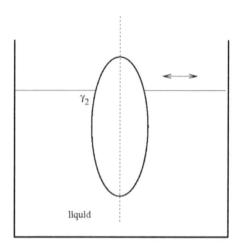

Fig. 5.11   Horizontal capillary surface, zero gravity

5.3   Appendix

In the following we prove formulas used in previous sections. Let X be the set of all $w = (\zeta, \alpha, a) \in C^1(U) \cap C^0(\overline{U}) \times \mathbb{R}^3 \times \mathbb{R}^3$ such that ζ satisfies the side conditions (5.3) and (5.4) on ∂U, i. e., ζ is proportional to $\overline{\nu}(0)$ along $\partial_1 S_0$ and $\partial_2 S_0$. See Fig. 5.4. Further let X_0 be the subset such that w satisfies the side conditions

$$\int_{S_0} \xi \, dA = 0$$

and

$$\int_{S_0} \chi(u)\langle N_{S_0}(u), (\alpha_i R_{\alpha_i}(0))x(u) + a\rangle \, dA$$
$$+ \int_{W_2(S_0)} \langle N_{\Sigma_0}(v), (\alpha_i R_{\alpha_i}(0))y(v) + a\rangle \, dA = 0.$$

We recall that

$$z_0(u, \epsilon) = x(u) + \epsilon\zeta(u) + r(u, \epsilon),$$
$$z(u, \mu, \epsilon) = (1 - \chi(u))z_0(u, \epsilon) + \chi(u)\left(R(\mu\alpha)z_0(u, \epsilon) + \mu a\right),$$
$$z^*(u, \mu, \epsilon, q) = z^*(u, \mu, \epsilon) + q\xi_0(u)N_0(u).$$

Lemma 5.1. *For given $w \in X_0$ there exists a regular function $q = q(c)$ such that the family of admissible configurations given by $(\mathcal{S}^*(c), \Sigma(c))$, where $\mathcal{S}^*(c)$ is defined through $z^*(u, c, q(c))$, is volume preserving for all $c \in \mathbb{R}^2$,*

$|c| < c_0$. *The function* $q(c)$ *satisfies* $q(0,0) = 0$, $q_\mu(0,0) = 0$ *and* $q_\epsilon(0,0) = 0$.

Proof. We have, see Fig. 5.3,

$$\Omega_l(\mathcal{S}^*(\mu,\epsilon,q), \Sigma(\mu,\epsilon)) = \Omega^1 \cup \Omega^2 \cup \Omega^3 \cup \Omega^4 \cup \Omega_l(\mathcal{S}_0, \Sigma_0),$$

where

$$\Omega^1 = \{z^*(u,\mu,t,0): u \in U, \ 0 < t < \epsilon\},$$
$$\Omega^2 = \{z^*(u,\rho,0,0): u \in U, \ 0 < \rho < \mu\},$$
$$\Omega^3 = \{z_\Sigma(v,\rho): v \in V_0, \ 0 < \rho < \mu\},$$
$$\Omega^4 = \{z^*(u,\mu,\epsilon,\tau): u \in U, \ 0 < \tau < q\}.$$

Here we assume for simplicity that $\mathcal{S}^*(\mu,\epsilon,q)$ and $\Sigma(\mu,\epsilon)$ are above of $\mathcal{S}_0$ and of Σ_0, resp. The final formula is valid for every $\mathcal{S}^*(\mu,\epsilon,q)$ and $\Sigma(\mu,\epsilon)$ we omit the details. For the general case see a remark in [Dierkes, Hildebrandt, Küster and Wohlrab (1992)], p. 81. We point out that the dashed surfaces in Fig. 5.3 are identically since supp $\chi \cap$ supp $\xi_0 = \emptyset$.
    Set

$$f(\mu,\epsilon,q) = |\Omega_l(\mathcal{S}^*(\mu,\epsilon,q), \Sigma(\mu,\epsilon))|.$$

From

$$\det \frac{\partial z^*(u,\mu,t,0)}{\partial(u,t)} = \langle z_\alpha^* \wedge z_\beta^*, z_t^* \rangle$$
$$= \langle N^*(u,\mu,t,0), z_t^*(u,\mu,t,0) \rangle W(u,\mu,t,0),$$

$$\det \frac{\partial z^*(u,\rho,0,0)}{\partial(u,\rho)} = \langle z_\alpha^* \wedge z_\beta^*, z_\rho^* \rangle$$
$$= \langle N^*(u,\rho,0,0), z_t^*(u,\rho,0,0) \rangle W(u,\rho,0,0),$$

$$\det \frac{\partial z_\Sigma(v,\rho)}{\partial(v,\rho)} = \langle z_{\Sigma,v_1} \wedge z_{\Sigma,v_2}, z_{\Sigma,\rho} \rangle$$
$$= \langle N_\Sigma(v,\rho), z_{\Sigma,\rho}(v,\rho) \rangle W_\Sigma(v,\rho),$$

$$\det \frac{\partial z^*(u,\mu,\epsilon,\tau)}{\partial(u,\tau)} = \langle z_\alpha^* \wedge z_\beta^*, z_\tau^* \rangle$$
$$= \langle N^*(u,\mu,\epsilon,\tau), z_\tau^*(u,\mu,\epsilon,\tau) \rangle W(u,\mu,\epsilon,\tau),$$

we find that

$$f(\mu, \epsilon, q) = \int_U \int_0^\epsilon \langle N^*(u, \mu, t, 0), z_t^*(u, \mu, t, 0)\rangle W(u, \mu, t, 0)\, dudt$$

$$+ \int_U \int_0^\mu \langle N^*(u, \rho, 0, 0), z_\rho^*(u, \rho, 0, 0)\rangle W(u, \rho, 0, 0)\, dud\rho$$

$$+ \int_U \int_0^q \langle N^*(u, \mu, \epsilon, \tau), z_\tau^*(u, \mu, \epsilon, \tau)\rangle W(u, \mu, \epsilon, \tau)\, dud\tau$$

$$+ \int_{V_0} \int_0^\mu \langle N_\Sigma(v, \rho), z_{\Sigma,\rho}(v, \rho)\rangle W_\Sigma(v, \rho)\, dvd\rho$$

$$+ |\Omega_l(S_0, \Sigma_0)|.$$

The assertion of the lemma follows since

$$f_q(0, 0, 0) = \int_{S_0} \xi_0\, dA = 1,$$

$$f_\epsilon(0,0,0) = \int_{S_0} \xi\, dA$$

and

$$f_\mu(0, 0, 0) = \int_{S_0} \chi(u)\langle N_{S_0}(u), (\alpha_i R_{\alpha_i}(0))x(u) + a\rangle\, dA$$

$$+ \int_{\mathcal{W}(S_0)} \langle N_{\Sigma_0}(v), (\alpha_i R_{\alpha_i}(0))y(v) + a\rangle\, dA.$$

$\square$

On the other hand we have

Corollary 5.1. *Suppose that for given $(\zeta, \alpha, a) \in X$ and a given regular scalar function $q(c)$, where $q(0, 0) = 0$, $q_\mu(0, 0) = 0$, $q_\epsilon(0, 0) = 0$, the configuration $(S^*(c), \Sigma(c))$ defined through $z^*(u, c) = z(u, c) + q(c)\xi_0(u)N_0(u)$, $u \in U$, and $z_\Sigma(v, \mu)$, $v \in V(\epsilon)$, is volume preserving, then (ζ, α, a) is in the subspace X_0.*

Proof. From the formulas in the proof above we see that $f_\mu(0, 0, 0) + f_q(0, 0, 0)q_\mu(0, 0, 0) = 0$ and $f_\epsilon(0, 0, 0) + f_q(0, 0, 0)q_\epsilon(0, 0, 0) = 0$. $\square$

We recall that $z(u, c) := (1 - \chi(u))z_0(u, \epsilon) + \chi(u)\left(R(\mu\alpha)z_0(u, \epsilon) + \mu a\right)$.

Lemma 5.2.

$$\frac{\partial}{\partial c_k}|S(c)| = -2\int_U H(u, c)\langle N(u, c), z_{c_k}(u, c)\rangle W(u, c)\, du$$

$$+ \int_{\partial_1 S(c)} \langle z_{c_k}(u, c), \nu(u, c)\rangle\, ds(c)$$

$$+ \int_{\partial_2 S(c)} \langle z_{c_k}(u, c), \nu(u, c)\rangle\, ds(c),$$

where $H(u,c)$ denotes the mean curvature of $\mathcal{S}(c)$ at $u \in U$, and

$$ds(c) = |Z_\tau(\tau, c)| d\tau, \quad Z(\tau, c) := z(u(\tau), c),$$

here $u(\tau)$ is a regular parameter representation of $\partial_1 U$ or of $\partial_2 U$, resp.

Proof. Set $u = (\alpha, \beta)$, then

$$\frac{\partial}{\partial c_k} |\mathcal{S}(c)| = \int_U \frac{1}{W(u,c)} \big(E(u,c) \langle z_\beta(u,c), z_{\beta,c_k}(u,c) \rangle$$
$$- F(u,c)[\langle z_\alpha(u,c), z_{\beta,c_k}(u,c) \rangle + \langle z_\beta(u,c), z_{\alpha,c_k}(u,c) \rangle]$$
$$+ G(u,c) \langle z_\alpha(u,c), z_{\alpha,c_k}(u,c) \rangle \big) \, du.$$

The formula of Lemma 5.1 follows by integration by parts, see [Dierkes, Hildebrandt, Küster and Wohlrab (1992); Dierkes, Hildebrandt, and Sauvigny (2010)], p. 45, p. 44, resp., and by using the formula $\triangle z = 2HN$, see [Dierkes, Hildebrandt, Küster and Wohlrab (1992); Dierkes, Hildebrandt, and Sauvigny (2010)], p. 71, p. 72, resp., for a proof of this formula. $\square$

Corollary 5.2.

$$\left[\frac{\partial}{\partial \mu} |\mathcal{S}(c)| \right]_{c=0} = -2 \int_{\mathcal{S}_0} H(u,0) \langle N(u,0), \chi(u) \left((\alpha_i R_{\alpha_i}(0)) x(u) + a \right) \rangle \, dA$$
$$+ \int_{\partial_2 \mathcal{S}_0} \langle \alpha_i R_{\alpha_i} x(u) + a, \nu(u,0) \rangle \, ds,$$

$$\left[\frac{\partial}{\partial \epsilon} |\mathcal{S}(c)| \right]_{c=0} = -2 \int_{\mathcal{S}_0} H(u,0)\xi(u) \, dA + \int_{\partial_1 \mathcal{S}_0} \eta(u) \, ds + \int_{\partial_2 \mathcal{S}_0} \eta(u) \, ds.$$

Proof. The first formula of the corollary is a consequence of the formula

$$z_\mu(u,0) = \chi(u) \left((\alpha_i R_{\alpha_i}(0)) x(u) + a \right).$$

We recall that $\chi(u) = 0$ if $u \in \partial_1 U$ and $\chi(u) = 1$ if $u \in \partial_2 U$. The second formula of the previous corollary follows since

$$z_\epsilon(u,0) = \zeta,$$

and $\zeta = \xi(u)N_{\mathcal{S}_0}(u) + \eta(u)T_{\mathcal{S}_0}(u)$. $\square$

Lemma 5.3.

$$\frac{\partial}{\partial \epsilon} |\mathcal{W}_k(\mathcal{S}(c))| = \int_{\partial_k U} \langle Z_{0,\epsilon}(\tau,\epsilon), \overline{\nu}(u,0,\epsilon) \rangle |Z_{0,\tau}(\tau,\epsilon)| \, d\tau,$$

$k = 1, 2$, where $Z_0(\tau, t) = z_0(u(\tau), t)$ and $u(\tau) = u_k(\tau)$ is a regular parameter representation of $\partial_1 U$ or $\partial_2 U$, resp., and

$$\frac{\partial}{\partial \mu} |\mathcal{W}_k(\mathcal{S}(c))| = 0, \quad k = 1, 2.$$

Proof. Since $z(u, \mu, \epsilon) = z(u, 0, \epsilon)$ on $\partial_1 U$ and $|\mathcal{W}_2(\mathcal{S}(0, \epsilon))|$ does not change under rigid motion we get

$$|\mathcal{W}_k(\mathcal{S}(\mu, \epsilon))| = |\mathcal{W}_k(\mathcal{S}(0, \epsilon))|.$$

Let $u(\tau)$ be a regular parameter representation of $\partial_1 U$ or $\partial_2 U$. Then

$$|\mathcal{W}_k(\mathcal{S}(0, \epsilon))| = \int_{\partial_k U} \int_0^\epsilon |Z_{0,\tau}(\tau, t) \wedge Z_{0,t}(\tau, t)| \, d\tau dt + |\mathcal{W}_k(\mathcal{S}(0,0))|.$$

Here we assume for simplicity that $\partial_1 \mathcal{S}(0, \epsilon)$ is above of $\partial_1 \mathcal{S}(0,0)$. The final formula is valid for every $\partial_1 \mathcal{S}(0, \epsilon)$, we omit the details. For the general case see a remark in [Dierkes, Hildebrandt, Küster and Wohlrab (1992)], p. 81. The assertion of the lemma follows since the tangential plane on the container wall at $P \in \partial_1 \mathcal{S}(0, t)$ is spanned by the orthogonal vectors $\bar{\nu}(u(\tau), 0, t)$ and $\mathbf{t}(\tau, t) = Z_{0,\tau}(\tau, t)/|Z_{0,\tau}(\tau, t)|$. We recall that $\bar{\nu}(u(\tau), 0, t)$ and $\mathbf{t}(\tau, t)$ are orthogonal vectors. Then $Z_{0,t}(\tau, t) = a\bar{\nu}(u(\tau), 0, t) + bt(\tau, t)$ with scalar functions a and b depending on τ and t. Using the Lagrange identity, we find that

$$
\begin{aligned}
|Z_{0,\tau}(\tau, t) \wedge Z_{0,t}(\tau, t)|^2 &= |Z_{0,\tau}(\tau, t)|^2 |Z_{0,t}(\tau, t)|^2 - \langle Z_{0,\tau}(\tau, t), Z_{0,t}(\tau, t) \rangle^2 \\
&= |Z_{0,\tau}(\tau, t)|^2 (a^2 + b^2) \\
&\quad - \langle Z_{0,\tau}(\tau, t), a\bar{\nu}(u(\tau), 0, t) + bt(\tau, t) \rangle^2 \\
&= |Z_{0,\tau}(\tau, t)|^2 (a^2 + b^2) - b^2 |Z_{0,\tau}(\tau, t)|^2 \\
&= a^2 |Z_{0,\tau}(\tau, t)|^2 \\
&= \langle Z_{0,t}(\tau, t), \bar{\nu}(u(\tau), 0, t) \rangle |Z_{0,\tau}(\tau, t)|^2
\end{aligned}
$$

$\square$

Corollary 5.3.

$$\left[\frac{\partial}{\partial \epsilon} |\mathcal{W}_k(\mathcal{S}(c))| \right]_{c=0} = \int_{\partial_k S_0} (\xi(u) \sin \gamma_k + \eta(u) \cos \gamma_k) \, ds.$$

Proof. The formula follows since $|Z_{0,\tau}(\tau, 0)| d\tau = ds$ and

$$
\begin{aligned}
\langle Z_{0,\epsilon}(\tau, 0), \bar{\nu}(u(\tau), 0, 0) \rangle &= \langle \zeta(u(\tau)), \bar{\nu}(u(\tau), 0, 0) \rangle \\
&= \xi(u(\tau)) \langle N(u(\tau), 0, 0), \bar{\nu}(u(\tau), 0, 0) \rangle \\
&\quad + \eta(u(\tau)) \langle \nu(u(\tau), 0, 0), \bar{\nu}(u(\tau), 0, 0) \rangle \\
&= \xi(u(\tau)) \sin \gamma_k + \eta(u(\tau)) \cos \gamma_k.
\end{aligned}
$$

$\square$

Remark 5.5. The vector ζ is proportional to $\overline{\nu}(u(\tau), 0, 0)$ on $\partial_k S_0$, i. e., $\zeta = a(\tau)\overline{\nu}(u(\tau), 0, 0)$, where $a(\tau)$ is a sufficiently regular scalar function.

Lemma 5.4.

$$\frac{\partial}{\partial \epsilon} \int_{\Omega_l(\mathcal{S}(c), \Sigma(c))} F_1(x) \, dx = \int_U F_1(z(u, c))\langle N(u, c), z_\epsilon(u, c)\rangle W(u, c) \, du,$$

$$\frac{\partial}{\partial \mu} \int_{\Omega_l(\mathcal{S}(c), \Sigma(c))} F_1(x) \, dx =$$

$$\int_U \int_0^\epsilon \frac{\partial}{\partial \mu} \Big[F_1(z(u, \mu, t))\langle N(u, \mu, t), z_t(u, \mu, t)\rangle W(u, \mu, t) \Big] \, dudt$$

$$+ \int_U F_1(z(u, \mu, 0))\langle N(u, \mu, 0), z_\mu(u, \mu, 0)\rangle W(u, \mu, 0) \, du$$

$$+ \int_{V_0} F_1(z_\Sigma(v, \mu))\langle N_\Sigma(v, \mu), z_{\Sigma, \mu}(v, \mu)\rangle W_\Sigma(v, \mu) \, dv.$$

Proof. Here we assume for simplicity that $\mathcal{S}(c)$ and $\Sigma(c)$ are above of $\mathcal{S}(0)$ and of $\Sigma(c)$, resp. The final formula is valid for every $\mathcal{S}(c)$ and $\Sigma(c)$, we omit the details. For the general case see a remark in [Dierkes, Hildebrandt, Küster and Wohlrab (1992)], p. 81. Then, see Fig. 5.12,

$$\Omega_l(\mathcal{S}(c), \Sigma(c)) = \Omega^1 \cup \Omega^2 \cup \Omega^3 \cup \Omega_l(\mathcal{S}(0), \Sigma(0)),$$

with

$$\Omega^1 = \{z(u, \mu, t) : \ u \in U, \ 0 < t < \epsilon\},$$
$$\Omega^2 = \{z(u, \rho, 0) : \ u \in U, \ 0 < \rho < \mu\},$$
$$\Omega^3 = \{z_\Sigma(v, \rho) : \ v \in V_0, \ 0 < \rho < \mu\}.$$

Consequently,

$$\int_{\Omega_l(\mathcal{S}(c), \Sigma(c))} F_1(x) \, dx = \int_U \int_0^\epsilon F_1(z(u, \mu, t)) \det \frac{\partial z(u, \mu, t)}{\partial(u, t)} \, dudt$$

$$+ \int_U \int_0^\mu F_1(z(u, \rho, 0)) \det \frac{\partial z(u, \rho, 0)}{\partial(u, \rho)} \, dud\rho$$

$$+ \int_{V_0} \int_0^\mu F_1(z_\Sigma(v, \rho)) \det \frac{\partial z_\Sigma(v, \rho)}{\partial(v, \rho)} \, dvd\rho$$

$$+ \int_{\Omega_l(\mathcal{S}(0), \Sigma(0))} F_1(x) \, dx.$$

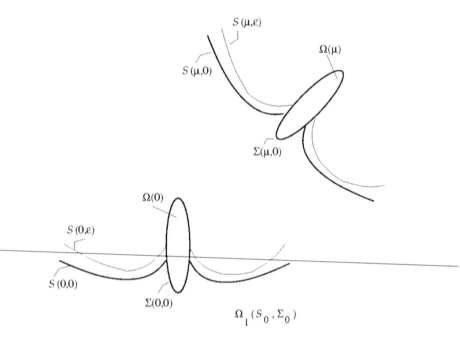

Fig. 5.12   Changing of the liquid region

Since

$$\det \frac{\partial z(u,\mu,t)}{\partial(u,t)} = \langle z_\alpha \wedge z_\beta, z_t \rangle$$

$$= \langle N(u,\mu,t), z_t(u,\mu,t) \rangle W(u,\mu,t),$$

$$\det \frac{\partial z(u,\rho,0)}{\partial(u,\rho)} = \langle z_\alpha \wedge z_\beta, z_\rho \rangle$$

$$= \langle N(u,\rho,0), z_\rho(u,\rho,0) \rangle W(u,\rho,0),$$

$$\det \frac{\partial z_\Sigma(v,\rho)}{\partial(v,\rho)} = \langle z_{\Sigma,v_1} \wedge z_{\Sigma,v_2}, z_{\Sigma,\rho} \rangle$$

$$= \langle N_\Sigma(v,\rho), z_{\Sigma,\rho}(v,\rho) \rangle W_\Sigma(v,\rho),$$

we obtain the formulas of the lemma.   □

Remark 5.6. Since the particle is moved by a rigid motion, we have $W_\Sigma(v,\mu) = W_\Sigma(v,0)$ and $N_\Sigma(v,\mu) = R(\mu\alpha)N_\Sigma(v,0)$.

Corollary 5.4.

$$\left[\frac{\partial}{\partial\epsilon}\int_{\Omega_l(\mathcal{S}(c),\Sigma(c))}F_1(x)\,dx\right]_{c=0}=\int_{\mathcal{S}_0}F_1(x(u))\xi(u)\,dA,$$

$$\left[\frac{\partial}{\partial\mu}\int_{\Omega_l(\mathcal{S}(c),\Sigma(c))}F_1(x)\,dx\right]_{c=0}=\int_{\mathcal{S}_0}F_1(x(u))\langle N(u,0,0),z_\mu(u,0,0)\rangle\,dA$$

$$+\int_{\mathcal{W}_2(\mathcal{S}_0)}F_1(y(v))\langle N_{\Sigma_0}(v,0),z_{\Sigma,\mu}(v,0)\rangle\,dA,$$

where

$$z_\mu(u,0,0)=\chi(u)\left((\alpha_iR_{\alpha_i}(0))x(u)+a\right),$$
$$z_{\Sigma,\mu}(v,0)=(\alpha_iR_{\alpha_i}(0))y(v)+a.$$

Lemma 5.5.

$$\frac{d}{d\mu}\int_{\Omega(\mu)}F_2(y)\,dy=\int_{\Omega(0)}\langle\nabla F_2(R(\mu\alpha)y+\mu a),(\alpha_iR_{\alpha_i}(\mu\alpha))y+a\rangle\,dy.$$

Proof. Since

$$\Omega(\mu)=\{q\in\mathbb{R}^3:\ q=R(\mu\alpha)y+\mu a,\ \text{where }y\in\Omega_0\},$$

we get

$$\int_{\Omega(\mu)}F_2(q)\,dq=\int_{\Omega(0)}F_2(R(\mu\alpha)y+\mu a)\left|\det\frac{\partial q}{\partial y}\right|\,dy$$

$$=\int_{\Omega(0)}F_2(R(\mu\alpha)y+\mu a)\,dy.$$

$\square$

Corollary 5.5.

$$\left[\frac{d}{d\mu}\int_{\Omega(\mu)}F_2(y)\,dy\right]_{c=0}=\int_{\Omega(0)}\langle\nabla F_2(y),(\alpha_iR_{\alpha_i}(0))y+a\rangle\,dy.$$

5.4 Problems

(1) Consider a floating body in an equilibrium which floats in a liquid of infinite volume and suppose that the capillary interface is horizontal, see Fig. 5.13. Show that $\mathcal{B}$ = weight of displaced fluid, where $\mathcal{B}$ denotes the buoyancy of the body.

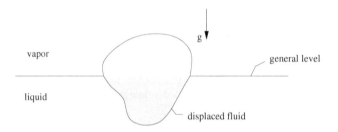

Fig. 5.13   Displaced liquid=buoyancy

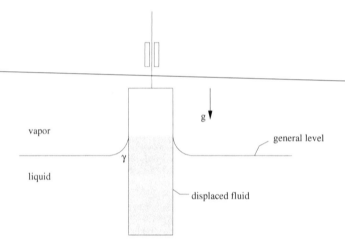

Fig. 5.14   Capillarity and buoyancy

(2) Consider a cylinder of constant circular cross section which hangs in a fluid of infinite volume. Let γ be the constant contact angle between the liquid and the body and suppose the cylinder can move vertically only, see Fig. 5.14. Show that

$$\mathcal{B} = \begin{cases} < \text{ weight of displaced fluid if } \gamma < \pi/2 \\ > \text{ weight of displaced fluid if } \gamma > \pi/2 \\ = \text{ weight of displaced fluid if } \gamma = \pi/2 \end{cases}.$$

The displaced fluid is sketched in Fig. 5.14.

(3) Determine $\lim_{d \to \infty} \mathcal{F}_x^L(d, \gamma_w, \gamma_l)$, see Sec. 4.2.1 (attraction/repelling of a plate).

(4) Express $\mathcal{F}_x^L(d, \gamma_w, \gamma_l)$ in terms of elliptic integrals.

(5) Discuss the sign of $\mathcal{F}_x^L(d, \gamma_w, \gamma_l)$.

(6) Discuss the rotation of a cylindrical body, with a constant square as cross section, around the middle axis. It is assumed that this axis coincides with the middle axis of the container which has a constant square as cross section, see Fig. 5.15.

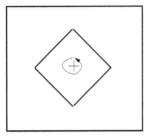

Fig. 5.15   Rotation in a tube

Chapter 6

Wetting barriers

Consider a container or a floating particle with edges or lines where different materials meet each other, see [Miersemann (2002)] for the case of wetting barriers on the container wall. Another type of a wetting barrier is the

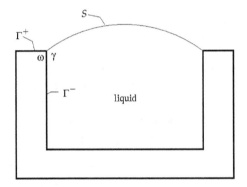

Fig. 6.1   A container with an edge, $\gamma^- \leq \gamma \leq \gamma^+ - \omega + \pi$

boundary of the liquid on a homogeneous container wall if one assumes that the adhesion coefficients of the wetted part and the non-wetted part of the wall are different. This assumption leads to inequalities for the contact angle.

Suppose that the capillary surface hangs on such a line of the container wall or the particle, see Fig. 6.1 and Fig. 6.2. A typical example of such a floating particle is a floating razor blade. Along the wetting barrier the contact angle is not determined, see a remark in [Gyemant (1927)], p. 356. Suppose that the configuration is in an equilibrium, see the definition below, then it turns out that the contact angle is in an interval, defined by the data. We will see that all the equilibrium conditions of Theorem 5.1 hold

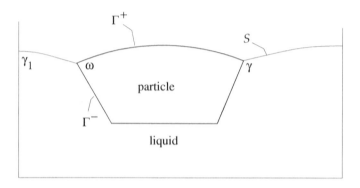

Fig. 6.2   A particle with an edge, $\gamma^- \le \gamma \le \gamma^+ - \omega + \pi$

with the only exception that we have to replace the equation which defines the contact angle by inequalities.

6.1   Edges

In the following we consider the case of a floating particle with an edge. The case of a container with an edge is contained in this case if one keeps the particle fixed, i. e., $a = 0$ and $\alpha = 0$, and if the boundary of the particle is replaced by the container wall. In contrast to the case of smooth particles, we form two families $\mathcal{S}_0^\pm(\epsilon)$ of admissible comparison surfaces by

$$z_0^\pm(u, \epsilon) = x(u) + \epsilon\zeta^\pm(u) + O(\epsilon^2),$$

where $0 < \epsilon < \epsilon_0$, and $\zeta^\pm = \xi^\pm(u)N_{\mathcal{S}_0}(u) + \eta^\pm(u)T_{\mathcal{S}_0}(u)$ are given. We recall that $\mathcal{S}_0^\pm(\epsilon)$ is called admissible if it meets Ω_0 at its boundary Σ_0 only. More precisely, we assume that $z_0^\pm(u, \epsilon) \in \mathbb{R}^3 \setminus \overline{\Omega}$ for all $u \in U$ and $0 < \epsilon \le \epsilon_0$, and $z_0^\pm(u, \epsilon) \in \Sigma_0^\pm$ if $u \in \partial U$, see Fig. 6.4 for notations.

Suppose that the interior open angle ω of the edge satisfies $0 < \omega < \pi$, then it can be shown, we omit the details, that $z_0^-(u, \epsilon)$ defines an admissible family of comparison surfaces if $0 < \gamma < \pi$, and $z_0^+(u, \epsilon)$ is admissible if $\pi - \omega < \gamma < 2\pi - \omega$, see Fig. 6.3 (a). If $\pi < \omega < 2\pi$, see Fig. 6.3 (b), then $z_0^+(u, \epsilon)$ and $z_0^-(u, \epsilon)$ define admissible comparison surfaces. Since $\zeta^+ \perp N_{\Sigma_0^+}$, $\zeta^- \perp N_{\Sigma_0^-}$, where $u \in \partial U_2$, see Fig. 6.4 for notations, then ζ^+ satisfies

$$\xi^+ \cos(\gamma + \omega) - \eta^+ \sin(\gamma + \omega) = 0 \tag{6.1}$$

at ∂U_2, and in the case of the second family we have

$$-\xi^- \cos\gamma + \eta^- \sin\gamma = 0 \tag{6.2}$$

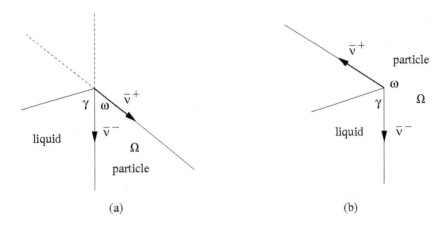

Fig. 6.3   Convex and reentrant edge

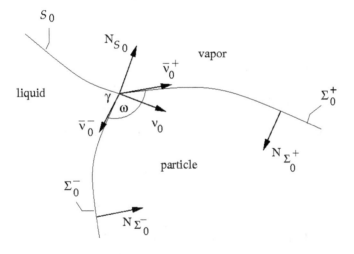

Fig. 6.4   Particle with an edge, notations

at $\partial_2 U$. Let $\mathcal{S}^{\pm}(c)$ be the comparison surface defined by

$$z^{\pm}(u, c) = z_0^{\pm}(u, \epsilon) + \chi(u)\left(R(\mu\alpha)z_0^{\pm}(u, \epsilon) + \mu a - z_0^{\pm}(u, \epsilon)\right),$$

where $c = (\mu, \epsilon)$, $\epsilon \geq 0$ and $|\mu| < \mu_0$.

Suppose that

$$\mathcal{E}(\mathcal{S}^{\pm}(c), \Sigma(c)) \geq \mathcal{E}(\mathcal{S}(0), \Sigma(0)), \quad |\Omega_l(\mathcal{S}^{\pm}(c), \Sigma(c))| = C.$$

In contrast to Chap. 5, one of the resulting equations has to be replaced by the inequality

$$\left[\frac{\partial}{\partial\epsilon}\mathcal{E}(\mathcal{S}^{\pm}(c),\Sigma(c))\right]_{c=0} \geq 0, \quad \left[\frac{\partial}{\partial\mu}\mathcal{E}(\mathcal{S}^{\pm}(c),\Sigma(c))\right]_{c=0} = 0. \qquad (6.3)$$

The second necessary condition is the equation

$$\left[\nabla_c|\Omega_l(\mathcal{S}^{\pm}(c),\Sigma(c))|\right]_{c=0} = 0. \qquad (6.4)$$

We call a capillary surface in an equilibrium if (6.3) and (6.4) are satisfied

Let $X^{\pm}$ be the set of all $(\zeta^{\pm},\alpha,a) \in C^1(U)\cap C^0(\overline{U}) \times \mathbb{R}^3 \times \mathbb{R}^3$ satisfying the angle conditions (6.1) and (6.2), resp., and $\zeta^{\pm} = a(u)\bar{\nu}^{\pm}$ on ∂U_2, where $a(u)$ is non-negative and continuous on ∂U_2. Thus X^+ and X^- define convex cones with the vertex at zero. From a Lagrange multiplier rule of Chap. 10 it follows that there is a real λ_0 such that

$$\left[\frac{\partial}{\partial\epsilon}\mathcal{L}(\mathcal{S}^{\pm}(c),\Sigma(c),\lambda_0)\right]_{c=0} \geq 0, \qquad (6.5)$$

$$\left[\frac{\partial}{\partial\mu}\mathcal{L}(\mathcal{S}^{\pm}(c),\Sigma(c),\lambda_0)\right]_{c=0} = 0 \qquad (6.6)$$

for all $(\zeta,\alpha,a) \in X^+$ or in X^-, resp. Following the proof of Theorem 5.1, we obtain the differential equation (5.13), the boundary condition (5.14) on the container wall and, see the proof of Theorem 5.1,

$$\left[\frac{\partial}{\partial\epsilon}\mathcal{L}(\mathcal{S}^+(c),\Sigma(c),\lambda_0)\right]_{c=0} = \int_{\partial S_0}\langle\zeta^+,\nu_0 - \beta_2\bar{\nu}_0^+\rangle\,ds$$

$$= \int_{\partial S_0}a(u)\langle\bar{\nu}_0^+,\nu_0 - \beta_2\bar{\nu}_0^+\rangle\,ds$$

$$= \int_{\partial S_0}a(u)\left(-\cos(\gamma+\omega) - \beta_2\right)\,ds \geq 0$$

for all $a(u)$ which are continuous and non-negative at ∂U_2. Thus, since $\beta_2 = \cos\gamma^+$, we have the inequality

$$\cos\gamma^+ \leq \cos(\gamma+\omega - \pi). \qquad (6.7)$$

For the comparison surfaces $\mathcal{S}^-$ we get

$$\left[\frac{\partial}{\partial\epsilon}\mathcal{L}(\mathcal{S}^-(c),\Sigma(c),\lambda_0)\right]_{c=0} = \int_{\partial S_0}\langle\zeta^-,\nu_0 + \beta_2\bar{\nu}_0^-\rangle\,ds$$

$$= \int_{\partial S_0}a(u)\langle\bar{\nu}_0^-,\nu_0 + \beta_2\bar{\nu}_0^-\rangle\,ds$$

$$= \int_{\partial S_0}a(u)\left(-\cos\gamma + \beta_2\right)\,ds \geq 0$$

for all $a(u)$ which are continuous and non-negative at ∂U_2. Thus

$$\cos \gamma \le \cos \gamma^-. \tag{6.8}$$

Then we get the following necessary conditions.

Case $0 \le \omega < \pi$. If $0 \le \gamma \le \pi$ and $\pi - \omega < \gamma < 2\pi - \omega$, then there are admissible families of comparison surfaces $\mathcal{S}^+(c)$ and $\mathcal{S}^-(c)$ and, see (6.8) and (6.7),

$$\gamma^- \le \gamma \le \gamma^+ - \omega + \pi.$$

If $0 < \gamma < \pi - \omega$, then there exists an admissible family $\mathcal{S}^-(c)$ and (6.8) implies that $\gamma^- \le \gamma$.
If $\pi < \gamma \le 2\pi - \omega$, then there is an admissible family $\mathcal{S}^+(c)$ which implies that $\gamma \le \gamma^+ - \omega + \pi$.

Case $\pi < \omega < 2\pi$. In this case there exist admissible families of comparison surfaces $\mathcal{S}^+(\epsilon)$ and $\mathcal{S}^-(\epsilon)$. Then

$$\gamma^- \le \gamma \le \gamma^+ - \omega + \pi.$$

Summarizing, we get

Theorem 6.1. *Let γ be the angle between Γ^- and the capillary surface in equilibrium, then*

$$\gamma^- \le \gamma \le \gamma^+ - \omega + \pi$$

at $\partial_2 \mathcal{S}_0$. Additionally, in any case we have that $\gamma^-, \gamma^+ \in [0, \pi]$, $\omega \in [0, 2\pi)$ and $0 \le \gamma \le 2\pi - \omega$.

Equations (5.16) and (5.17) follow again from equation (6.6).

Remark 6.1. A capillary surface which hangs at a wetting barrier is sometimes called pinned at this line. A mathematical founded proof for this behavior of a liquid does not seem to appear in the literature with the exception of the paper [Miersemann (2002)]. If the strict inequalities

$$\gamma^- < \gamma < \gamma^+ - \omega + \pi$$

hold and if the admissible variations do not coincide with the equilibrium surface at the barrier line then this variation is positive. Moreover, an equilibrium defines a strict weak local minimizer of the energy provided the strict inequalities hold and if an additional eigenvalue criterion is satisfied, see [Miersemann (2002)], p. 246.

6.2　Stability

In the following we consider the case of a container with wetting barriers in the absence of a floating particle. Let $\mathcal{S}^\pm(\epsilon)$ be the family of volume preserving admissible comparison surfaces defined above through $z^\pm(u, \epsilon)$ where $a = 0$ and $\alpha = 0$ and suppose that $\mathcal{S}_0 = \mathcal{S}(0)$ is an equilibrium interface. Then

$$\mathcal{E}(\mathcal{S}^\pm(\epsilon)) - \mathcal{E}(\mathcal{S}_0) = \mathcal{L}(\mathcal{S}^\pm(\epsilon), \lambda) - \mathcal{L}(\mathcal{S}_0, \lambda)$$

$$= \epsilon \left[\frac{d}{d\epsilon} \mathcal{L}^\pm(\mathcal{S}(\epsilon), \lambda) \right]_{\epsilon=0} + \frac{\epsilon^2}{2} \left[\frac{d^2}{d\epsilon^2} \mathcal{L}^\pm(\mathcal{S}(\epsilon), \lambda) \right]_{\epsilon=0}$$

$$+ O(\epsilon^3).$$

We have

$$\left[\frac{d}{d\epsilon} \mathcal{L}^+(\mathcal{S}(\epsilon), \lambda) \right]_{\epsilon=0} = \int_{\partial\mathcal{S}_0} a(u)(\cos(\gamma + \omega - \pi) - \cos\gamma^+) \, ds,$$

$$\left[\frac{d}{d\epsilon} \mathcal{L}^-(\mathcal{S}(\epsilon), \lambda) \right]_{\epsilon=0} = \int_{\partial\mathcal{S}_0} a(u)(-\cos\gamma + \cos\gamma^-) \, ds.$$

Since $\zeta^\pm = \xi^\pm N_{\mathcal{S}_0} + \eta^\pm T_{\mathcal{S}_0}$ and $a > 0$ on a part of ∂U if and only if $\xi^\pm > 0$ on that part of ∂U provided $\xi^\pm \in C^0(\partial U)$ and γ is different from 0 or $2\pi - \omega$. Consequently, there is a positive constant $\epsilon_0(\zeta)$ such that

$$\mathcal{E}(\mathcal{S}^\pm(\epsilon)) - \mathcal{E}(\mathcal{S}_0) \geq 0$$

for all $0 < \epsilon < \epsilon_0(\zeta)$ if $\xi^+ > 0$, $\xi^- > 0$, resp., on a part of ∂U and if the strict inequalities $\gamma^- < \gamma < \gamma^+ - \omega + \pi$ are satisfied. Thus, see the corollary to Lemma 3.1 of Chapter 3, a necessary condition such that the extremal $\mathcal{S}$ defines a (weak or strong) local minimizer of the associated energy is the inequality

$$\sigma \int_{\mathcal{S}_0} \left(|\nabla\xi|^2 - 2(2H_0^2 - K_0)\xi^2 \right) \, dA + \int_{\mathcal{S}_0} \langle \nabla F_1, N_0 \rangle \xi^2 \, dA \geq 0$$

for all $\xi \in W_0^{1,2}(\mathcal{S}_0)$ satisfying the side condition $\int_{\mathcal{S}_0} \xi \, dA = 0$.

To answer the question whether a given extremal defines a strong minimizer of the associated energy functional requires the construction of an embedding family of configurations, see Chap. 2 and Chap. 3 for the case of liquid layers and capillary interfaces, resp.

6.3 Problems

(1) Consider a liquid which hangs at the upper edge of a cylindrical container with constant circular cross section, see Fig. 6.1. Suppose that $\omega = \pi/2$ and that the gravity vanishes. Show that the following stability criterion holds:

$$\int_{\mathcal{S}_0} \left(|\nabla \xi|^2 - 2(2H_0^2 - K_0)\xi^2 \right) \, dA > 0$$

for all $\xi \in W_0^{1,2}(\mathcal{S}_0) \setminus \{0\}$ satisfying the side condition $\int_{\mathcal{S}_0} \xi \, dA = 0$. *Hint:* $\mathcal{S}_0$ is a spherical cap, and see [Wente (1980)].

(2) Consider a floating razor blade. Find the necessary equilibrium conditions for γ.

(3) Consider a liquid trapped in a groove as shown in Fig. 6.5.

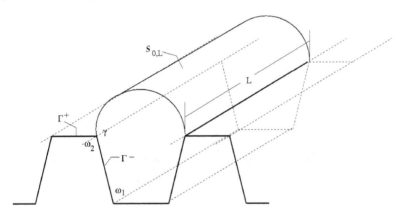

Fig. 6.5 Liquid trapped in a groove

(i) Find the necessary equilibrium conditions for γ.

(ii) Consider the case of zero gravity and check whether the following stability criterion holds:

$$\int_{\mathcal{S}_{0,L}} \left(|\nabla \xi|^2 - 2(2H_0^2 - K_0)\xi^2 \right) \, dA > 0$$

for all $\xi \in W_0^{1,2}(\mathcal{S}_0) \setminus \{0\}$. Here $\mathcal{S}_{0,L}$ denotes a section of length L of $\mathcal{S}_0$. *Hint:* $\mathcal{S}_0$ is a cylindrical cap.

(4) Find the volume of the liquid which rests at the lower end of a narrow tube, see Fig. 1.17 of the introduction.
(5) Discuss the regularity of the trace of the liquid which hangs at the upper edge of a wedge, see Fig. 1.16 of the introduction. The conjecture is that the trace is a $C^{1,\alpha}$-function.
(6) Find the stability bound of the radius of the dry spot of the drop with a hole, see Fig. 1.15 of the introduction.
(7) Discuss the case of a rounded upper edge of a tube with constant circular cross section, see [Miersemann (2002)], pp. 249–250, and Fig. 6.6, where Γ_ϵ is a circular arc of radius ϵ.

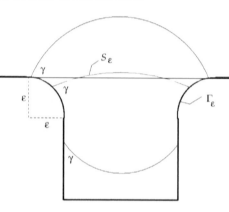

Fig. 6.6   Rounded edge

Chapter 7

Asymptotic formulas

Here we consider capillarity problems depending on a small or large real parameter. In general, there are no explicit solutions with the exception of few examples, see the previous sections. Therefore asymptotic formulas are of interest if the related parameter is small or large, resp. The main tools to get asymptotic formulas are comparison principles.

7.1 Comparison principles

The following comparison principles exploit the special strong non-linearity of the governing equation. There are no counterparts for linear elliptic equations, see some examples in this section. Set

$$Tu = \frac{\nabla u}{\sqrt{1 + |\nabla u|^2}} \quad \text{and} \quad Nu = \text{div } Tu - \kappa u,$$

where κ is a non-negative constant.

We recall that an open set is said to be connected if it is not a union of two nonempty open sets with empty intersection. Concerning the definition of the Hausdorff measure mentioned in the following theorem see for example [Evans and Gariepy (1992)].

Theorem 7.1 (Concus and Finn, 1970). *Let $\Omega \subset \mathbb{R}^2$ be a bounded connected domain and assume that the boundary $\Sigma = \partial\Omega$ admits a decomposition*

$$\Sigma = \Sigma_\alpha \cup \Sigma_\beta \cup \Sigma_0,$$

where

(a) Σ_0 can be covered, for any $\varepsilon > 0$, by a countable number of disks B_{δ_i} of radius δ_i, such that $\sum \delta_i < \varepsilon$, that is, $H^1(\Sigma_0) = 0$, where H^1 is the one-dimensional Hausdorff measure,

(b) Σ_α is Lipschitz and $\Sigma_\beta \in C^1$.

Let $u, v \in C^2(\Omega)$ with the properties

(i) $Nu \geq Nv$ in Ω,

(ii) $\limsup(u - v) \leq 0$ for any approach to Σ_α from Ω,

(iii) $(Tu - Tv) \cdot \nu \leq 0$ almost everywhere (with respect to $L^1(\Sigma_\beta)$) on Σ_β as a limit from points of Ω, where ν denotes the exterior unit normal at Σ_β.

(A) If $\kappa > 0$ or if $\Sigma_\alpha \neq \emptyset$, then $v \geq u$ in Ω. If $v(x_0) = u(x_0)$ at any point $x_0 \in \Omega$, then $v(x) \equiv u(x)$.

(B) If $\kappa = 0$ and $\Sigma_\alpha = \emptyset$, then $v(x) \equiv u(x) + const.$ in Ω.

Proof. See [Finn (1986)], p. 110. Define for a positive constant M

$$\Omega_0^M = \{x \in \Omega : \ 0 < u(x) - v(x) < M\},$$
$$\Omega^- = \{x \in \Omega : \ u(x) - v(x) \leq 0\},$$
$$\Omega^+ = \{x \in \Omega : \ u(x) - v(x) \geq M\},$$
$$\Lambda(\epsilon) = \Omega \cap \partial(\cup B_{\delta_i}),$$
$$\Omega(\epsilon) = \Omega \setminus \cup B_{\delta_i}.$$

Assume that $u > v$ on a subset of Ω. Then there exists a positive constant M such that $\Omega_0^M \cap \Omega(\epsilon) \neq \emptyset$ for all $0 < \epsilon < \epsilon_0$, ϵ_0 sufficiently small. Define for $x \in \Omega$ the function

$$w = \begin{cases} 0 & : & x \in \Omega^- \\ u - v & : & x \in \Omega_0^M \\ M & : & x \in \Omega^+ \end{cases}.$$

The function w belongs to the Sobolev class $W^{1,2}(\Omega) \cap L^\infty(\Omega)$, see for example [Gilbarg and Trudinger (1983)], p. 151. Since $w \geq 0$ and $Nu - Nv \geq 0$, it follows from the definition of N and after integration by parts that

$$0 \leq \int_{\Omega(\epsilon)} w(Nu - Nv) \, dx = -\int_{\Omega(\epsilon)} \nabla w \cdot (Tu - Tv) \, dx$$
$$- \kappa \int_{\Omega(\epsilon)} w(u - v) \, dx + \int_{\Lambda(\epsilon)} w(Tu - Tv) \cdot \nu \, dS$$
$$+ M \int_{\Sigma^+(\epsilon)} (Tu - Tv) \cdot \nu \, dS + \int_{\Sigma_0^M(\epsilon)} w(Tu - Tv) \cdot \nu \, dS,$$

where

$$\Sigma_0^M(\varepsilon) = \partial(\Omega_0^M \setminus \cap B_{\delta_i}) \cap \partial\Omega,$$
$$\Sigma^+(\varepsilon) = \partial(\Omega^+ \setminus \cap B_{\delta_i}) \cap \partial\Omega,$$
$$\Sigma^-(\varepsilon) = \partial(\Omega^- \setminus \cap B_{\delta_i}) \cap \partial\Omega,$$

see Fig. 7.1. Thus, assigning symbols to the integrals on the right in order

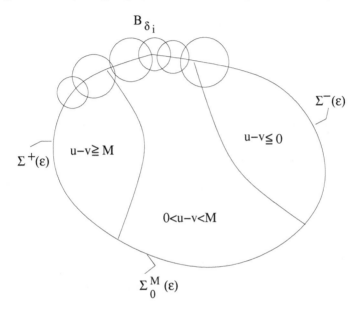

Fig. 7.1   Proof of Theorem 7.1

of appearance,

$$0 \le -Q(\varepsilon) - W(\varepsilon) + J_\Lambda(\varepsilon) + J^+(\varepsilon) + J_0^M(\varepsilon).$$

We observe that $W(\varepsilon) \ge 0$ and $W(\varepsilon) = 0$ if and only if $\kappa = 0$. Since $\Sigma^+(\varepsilon) \subset \Sigma_\beta$ it follows $J^+(\varepsilon) \le 0$ and since all points from $\Sigma_0^M(\varepsilon)$ where $w \ne 0$ are contained in Σ_β we have $J_0^M(\varepsilon) \le 0$. From the definition of w we see that

$$Q(\varepsilon) = \int_{\Omega_0^M(\varepsilon)} \nabla(u - v) \cdot (Tu - Tv)\, dx,$$

with $\Omega_0^M(\varepsilon) = \Omega_0^M \setminus (\cup B_{\delta_i})$. Thus $Q(\varepsilon) \ge 0$ and $Q(\varepsilon) = 0$ if and only if $\nabla u \equiv \nabla v$ in $\Omega_0^M(\varepsilon)$. This follows since the vector field $A(p) := p/\sqrt{1 + |p|^2}$, $p \in \mathbb{R}^n$, satisfies

$$\langle A(p) - A(q), p - q \rangle \ge c|p - q|^2,$$

where $c = c(p, q)$ is a positive constant, see an exercise.

Finally the inequality $|J_\Lambda(\varepsilon)| \leq 4\pi M\varepsilon$ follows from the fact that $|Tf| \leq 1$ for all $f \in C^1(\Omega)$.

Consequently,

$$\lim_{\varepsilon \to 0}(Q(\varepsilon) + W(\varepsilon)) \leq 0.$$

This implies, since $Q(\varepsilon)$ and $W(\varepsilon)$ are non-negative and increasing with decreasing ε, that $Q(\varepsilon) = 0$ and $W(\varepsilon) = 0$ for each ε, $0 < \varepsilon \leq \varepsilon_0$.

We conclude from $W(\varepsilon) = 0$ that in the case $\kappa > 0$ a construction of Ω_0^M is not possible for each ε and M. Thus $v \geq u$ in Ω.

Let $\kappa = 0$. If there is a subset where $v(x) < u(x)$, then $u - v \equiv c$, where c is a constant and $0 < c < M$ on each connected set of $\Omega_0^M(\varepsilon)$ since $\nabla u \equiv \nabla v$ on these sets according to the above considerations. The constants may differ from set to set. Since $u - v = 0$ or $u - v = M$ on the boundary of these components if the boundary point in consideration is in Ω and since $u - v$ is continuous in Ω by assumption. It follows that the union of these connected sets is Ω, see Fig. 7.2. Then we have $u - v = const. > 0$ in Ω.

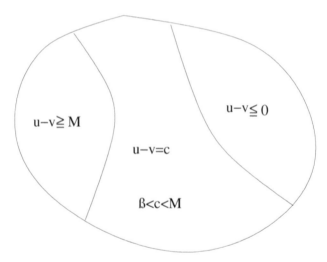

Fig. 7.2   Proof of Theorem 7.1

Case $\Sigma \neq \emptyset$. The above inequality is a contradiction to the assumption $\limsup(u - v) \leq 0$ from any approach to Σ_α from Ω. Consequently, $v \leq u$ in Ω.

Case $\Sigma = \emptyset$. In this case an arbitrary constant can be added to v without changing the hypotheses. Thus there is always a domain in Ω where $0 < u - (v - C) < M$. From above it follows that $u = v - C + const.$ in Ω.

It remains to show that from $v(x) \geq u(x)$ in Ω and $u(x_0) = v(x_0)$ for at least one $x_0 \in \Omega$ it follows that $u \equiv v$ in Ω. This is a consequence of the strong maximum principle of [Hopf (1952)][1] since

$$Nu - Nv = \sum_{i,j=1}^{2} a_{ij}(x) w_{x_i x_j}(x) + \sum_{i=1}^{2} b_i(x) w_{x_i}(x) - \kappa w,$$

where $w = u - v$, $a_{ij} = a_{ji}$, a_{ij}, $b_i \in C(\Omega)$, and the right hand side is elliptic in Ω. □

Remark 7.1. The previous result fails in the case of *linear* elliptic equations. In that case we can not take away in the boundary condition one point from the boundary as the following example shows. Let $\Omega \subset \mathbb{R}^2$ be the domain

$$\Omega = \{x \in B_1(0) : x_2 > 0\},$$

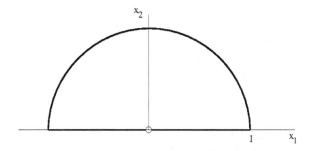

Fig. 7.3   Counterexample for linear problems

Assume that $u \in C^2(\Omega) \cap C(\overline{\Omega} \setminus \{0\})$ is a solution of

$$\triangle u = 0 \quad \text{in } \Omega,$$
$$u = 0 \quad \text{on } \partial\Omega \setminus \{0\}.$$

---

[1] *Strong maximum principle.* Suppose that $\Omega \subset \mathbb{R}^n$ is a connected domain. Let

$$Lu = \sum_{i,j=1}^{n} a_{ij}(x) w_{x_i x_j}(x) + \sum_{i=1}^{n} b_i(x) w_{x_i}(x) + c(x)w,$$

where $a_{i,j} = a_{ji}$, a_{ij}, b_i, $c \in C(\Omega)$, $c \leq 0$ in $\Omega \subset \mathbb{R}^n$, and the right hand side is elliptic in Ω. Assume that $w \in C^2(\Omega)$ satisfies $Lw \geq 0$ ($Lw \leq 0$) in Ω. Then if w achieves its positive supremum (negative infimum) in Ω, w is a constant.

This problem has solutions $u \equiv 0$ *and* $u = \mathrm{Im}(z + z^{-1})$, where $z = x_1 + ix_2$.

In contrast to this behavior of the Laplace equation, one has uniqueness according to the above theorem if $\triangle u = 0$ is replaced by the minimal surface equation

$$\frac{\partial}{\partial x_1}\left(\frac{u_{x_1}}{\sqrt{1 + |\nabla u|^2}}\right) + \frac{\partial}{\partial x_2}\left(\frac{u_{x_2}}{\sqrt{1 + |\nabla u|^2}}\right) = 0.$$

The following theorem concerns, in particular, unbounded domains.

Theorem 7.2 (Finn and Hwang, 1989). *Assume that* $\Omega \subset \mathbb{R}^2$, *not necessarily bounded,* $\kappa > 0$ *and that the assumptions (a)–(b) and (i)–(iii) of Theorem 7.1 are satisfied, then* $u \leq v$ *in* Ω. *If* $u(x) = v(x)$ *at any* $x \in \Omega$, *then* $u(x) \equiv v(x)$ *in* Ω.

Proof. If $u > v$ at some point of Ω, then there are positive constants m_1, m_2 and a set $\Omega_{12} \subset \Omega$ of positive measure where

$$0 < m_1 < u - v < m_2 < \infty.$$

Set

$$w(x) = \begin{cases} 0 & : & u - v \leq m_1 \\ u - v - m_1 & : & x \in \Omega_{12} \\ m_2 - m_1 & : & u - v \geq m_2 \, . \end{cases}$$

For any $R > 0$ set $B_R = \{x \in \mathbb{R}^n : |x| < R\}$ and

$$\Omega_R = \Omega \cap B_R, \quad \Gamma_R = \Omega \cap \partial B_R, \quad \Sigma_{\alpha,R} = \Sigma_\alpha \cap B_R, \quad \Sigma_{\beta,R} = \Sigma_\beta \cap B_R,$$

see Fig. 7.4. If R is sufficiently large, then the measure of $\Omega_R \cap \Omega_{12}$ is positive. Integration by parts gives

$$\int_{\Omega_R} w^{n-1}(\mathrm{div}\ Tu - \mathrm{div}\ Tv)\ dx$$

$$= \int_{\Gamma_R} w^{n-1}(Tu - Tv) \cdot \nu\ dS + \int_{\Sigma_{R,\beta}} w^{n-1}(Tu - Tv) \cdot \nu\ dS$$

$$+ \int_{\Sigma_{R,\alpha}} w^{n-1}(Tu - Tv) \cdot \nu\ dS - (n-1)\int_{\Omega_R} w^{n-2}(Tu - Tv) \cdot \nabla w\ dx.$$

Since $|(Tu - Tv) \cdot \nu| \leq 2$, $(Tu - Tv) \cdot \nu \leq 0$ on Σ_β, $w = 0$ on Σ_α and $(Tu - Tv) \cdot \nabla w \geq 0$, we find that

$$\kappa \int_{\Omega_R} w^{n-1}(u - v)\ dx \leq 2 \int_{\Gamma_R} w^{n-1}\ dS.$$

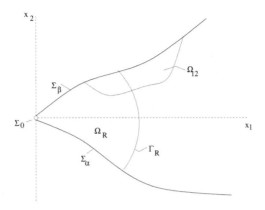

Fig. 7.4   Proof of Theorem 7.2

Since

$$\int_{\Omega_R} w^{n-1}(u-v)\,dx = \int_{\{u-v\le m_1\}} 0\,(u-v)\,dx$$

$$+ \int_{\{m_1<u-v<m_2\}} (u-v-m_1)^{n-1}(u-v)\,dx + \int_{\{u-v\ge m_2\}} (m_2-m_1)^{n-1}(u-v)\,dx$$

$$= \int_{\{u-v\le m_1\}} w^n\,dx + \int_{\{m_1<u-v<m_2\}} (u-v-m_1)^{n-1}(u-v-m_1)\,dx$$

$$+ m_1 \int_{\{m_1<u-v<m_2\}} (u-v-m_1)^{n-1}\,dx + \int_{\{u-v\ge m_2\}} (m_2-m_1)^{n-1}(m_2-m_1)\,dx$$

$$+ \int_{\{u-v\ge m_2\}} (m_2-m_1)^{n-1}((u-v)-(m_2-m_1))\,dx \ge \int_{\Omega_R} w^n\,dx.$$

we have

$$\int_{\Omega_R} w^n\,dx \le \int_{\Omega_R} w^{n-1}\,dx.$$

Thus we get finally

$$\kappa \int_{\Omega_R} w^n\,dx \le 2 \int_{\Gamma_R} w^{n-1}\,dS.$$

Set

$$Q(R) = \int_{\Omega_R} w^n\,dx.$$

Then

$$Q(R) \le \frac{2}{\kappa} \int_{\Gamma_R} w^{n-1}\,dS$$

$$\le \frac{2}{\kappa} \left(\int_{\Gamma_R} w^n\,dS \right)^{(n-1)/n} \left(\int_{\Gamma_R} dS \right)^{1/n}.$$

Consequently,

$$\frac{1}{\rho}Q(\rho)^{n/(n-1)} \le \left(\frac{2}{\kappa}\right)^{n/(n-1)} \frac{1}{\rho}\int_{\Gamma_\rho} w^n \, dS \left(\int_{\Gamma_\rho} dS\right)^{1/(n-1)}.$$

Set

$$J(R) = \int_{R_1}^R \frac{1}{\rho}Q(\rho)^{n/(n-1)} \, d\rho.$$

Let $0 < R_1 < R < \infty$, then, since $dS = \rho^{n-1}d\omega$, we obtain

$$J(R) \le \left(\frac{2}{\kappa}\right)^{n/(n-1)} \int_{R_1}^R \left(\frac{1}{\rho}\int_{\Gamma_\rho} w^n \, dS \left(\int_{\Gamma_\rho} dS\right)^{1/(n-1)}\right) d\rho$$

$$= \left(\frac{2}{\kappa}\right)^{n/(n-1)} \int_{R_1}^R \left(\int_{\Gamma_\rho} w^n \, dS \left(\int_{\Gamma_\rho/\rho} d\omega\right)^{1/(n-1)}\right) d\rho$$

$$\le |\omega_n|^{1/(n-1)} \left(\frac{2}{\kappa}\right)^{n/(n-1)} \int_{R_1}^R \int_{\Gamma_\rho} w^n \, dS \, d\rho$$

$$= |\omega_n|^{1/(n-1)} \left(\frac{2}{\kappa}\right)^{n/(n-1)} (Q(R) - Q(R_1)),$$

where ω_n denotes the n-dimensional unit sphere. Set

$$C_1 = |\omega_n|^{1/(n-1)} \left(\frac{2}{\kappa}\right)^{n/(n-1)},$$

then, if R_1 is sufficiently small, it follows that for all $R > R_1$

$$\frac{J'(R)}{(J(R))^{n/(n-1)}} \ge \frac{(Q(R))^{n/(n-1)}/R}{[C_1\,(Q(R) - Q(R_1))]^{n/(n-1)}}$$

$$\ge \frac{C_2}{R},$$

where C_2 is a positive constant which does not depend on R. Let $R_1 \le A < B < \infty$. Then

$$\int_A^B \frac{J'(\rho)}{(J(\rho))^{n/(n-1)}} \, d\rho \ge C_2(\ln B - \ln A)$$

or

$$-(n-1)\left[(J(A))^{-1/(n-1)} - (J(B))^{-1/(n-1)}\right] \ge C_2(\ln B - \ln A),$$

which is a contradiction since $0 < J(A) < J(B)$. The remainder of the proof follows from the strong maximum principle. $\qquad\square$

where $B = \kappa a^2$ is the Bond number. Extending a result of [Siegel (1980)] to higher order approximation of the ascent in a tube, we have shown in [Miersemann (1994)] the following result.

Proposition 7.2. *For every non-negative integer n there exist $n+1$ radially symmetric functions $\phi_l(y; \gamma)$, $l = 0, \ldots, n$, analytical in $B_1(0)$ and bounded on $\overline{B_1(0)}$, such that*

$$V(y; \gamma, B) = \frac{2 \cos \gamma}{B} + \sum_{l=0}^{n} \phi_l(y; \gamma) B^l + O(B^{n+1})$$

uniformly in $B_1(0)$ and γ, $0 \leq \gamma \leq \pi$, as B tends to zero. The function ϕ_0 is a solution of a nonlinear boundary value problem and the functions ϕ_l, $l \geq 1$ are solutions of linear boundary value problems. In particular, we find that

$$\phi_0(y; \gamma) = \frac{2}{3} \frac{1 - \sin^3 \gamma}{\cos^3 \gamma} - \frac{1}{\cos \gamma} \sqrt{1 - r^2 \cos^2 \gamma},$$

where $r = \sqrt{y_1^2 + y_2^2}$, and

$$\phi_1(y; \gamma) = -\frac{(1 - \sin \gamma)^2 (1 + 3 \sin^2 \gamma)}{6 \cos^7 \gamma} - \frac{1}{2 \cos^3 \gamma} \ln(1 + \sin \gamma)$$
$$+ \frac{(1 - \sin \gamma) \sin^2 \gamma}{3 \cos^5 \gamma} \frac{1}{\sqrt{1 - r^2 \cos^2 \gamma}}$$
$$+ \frac{1}{3 \cos^2 \gamma} \ln \left(1 + \sqrt{1 - r^2 \cos^2 \gamma} \right).$$

Remark 7.6. In the case $n = 0$ we have
$$V(y; \gamma, B) = \frac{2 \cos \gamma}{B} + \phi_0(y; \gamma) + O(B),$$
Then the center height of U is given by the *Laplace formula*

$$U(0) = \frac{2 \cos \gamma}{\kappa a} - \left(\frac{1}{\cos \gamma} - \frac{2}{3} \frac{1 - \sin^3 \gamma}{\cos^3 \gamma} \right) a + O(\kappa a^3),$$

see [Laplace (1806)] for a formal proof. The first mathematical proof of this formula was given in [Siegel (1980)].

Set $\beta = \cos \gamma$. In preparation of the proof of the above proposition we show the following asymptotic formulas.

Lemma 7.1. *For given $N \in \mathbb{N} \cup \{0\}$ there exist functions $\phi_l(r; \beta)$, $|\beta| \leq 1$, which are in $C^1[0, 1]$ and analytical in $0 \leq r < 1$ with respect to r and satisfy*

$$\sup |\phi_l(r; \beta)| \leq c_l < \infty$$

uniformly with respect to $|\beta| \le 1$ and $0 \le s \le 1$. And that

$$V_N := \frac{2\cos\gamma}{B} + \sum_{l=0}^{N} \phi_l B^l$$

satisfies

$$\operatorname{div} TV_N - BV_N = O(B^{N+1})$$

uniformly in $\Omega = B_1(0)$ and $|\beta| \le 1$, and

$$\lim_{s \to 1-0} \frac{V'_N}{\sqrt{1 + V_N'^2}} = \beta, \qquad V'_N(0) = 0,$$

where the derivatives are taken with respect to r.

Proof. We have

$$\operatorname{div} Tv_N - Bv_N = \frac{1}{r}\left(\frac{v'_N}{\sqrt{1 + (v'_N)^2}}\right)' - Bv_N.$$

The formal ansatz

$$V_N = \frac{C}{B} + \sum_{k=0}^{N} \phi_k B^k,$$

where C is a constant, and the conditions

$$\operatorname{div} TV_N - BV_N = \frac{1}{r}\left(\frac{V'_N}{\sqrt{1 + (V'_N)^2}}\right)' - BV_N = O(B^{N+1}), \quad (7.3)$$

$$\lim_{r \to 0} \frac{V'_N}{\sqrt{1 + (V'_N)^2}} = \beta \qquad\qquad (7.4)$$

lead to a nonlinear boundary value problem which determines ϕ_0 and to linear boundary value problems for the functions ϕ_k, $k \ge 1$. Let $\gamma \in [0, \pi/2]$. Then

$$\lim_{r \to 1-0} \frac{\phi'_0}{\sqrt{1 + \phi_0'^2}} = \cos\gamma,$$

$$\frac{(1 - \epsilon_0)\cos\gamma}{\sqrt{1 - r^2\cos^2\gamma}} \le \phi'_0(r) \le \frac{\cos\gamma}{\sqrt{1 - r^2\cos^2\gamma}}$$

for an $\epsilon_0 \in (0, 1)$ and for all $r \in (\epsilon_0, 1)$, and

$$0 \le (1 - r^2\cos^2\gamma)\phi''_0(r) \le \cos\gamma.$$

The functions ϕ_l satisfy

$$\frac{\phi_l'(r)}{(1+\phi_0(r)'^2)^{3/2}} = O(1-r)$$

as $r \to 1$ and uniformly in $\gamma \in [0, \pi/2]$, and

$$\sup |\phi_l(r; \gamma)| < \infty,$$

$$\sup \left((1 - r\cos\gamma)^{1/2} |\phi_l'(r; \gamma)| \right) < \infty,$$

$$\sup \left((1 - r\cos\gamma)^{3/2} |\phi_l''(r; \gamma)| \right) < \infty,$$

where the supremum is taken over $0 < r < 1$ and $0 \le \gamma \le \pi/2$. From the behavior of ϕ_0 it follows the asserted behavior of the functions ϕ_k, $k \ge 1$, using the strong non-linearity of the problem, see [Miersemann (1994)] for details. Since

$$V_N = \frac{C}{B} + \phi_0 + \sum_{k=1}^{N} \phi_k B^k,$$

we have

$$1 + V_N'^2 = (1 + \phi_0'^2)(1 + Q),$$

where

$$Q = \frac{2\phi_0' \sum_{k=1}^{N} \phi_k' B^k + \left(\sum_{k=1}^{N} \phi_k' B_k \right)^2}{1 + \phi_0'^2} \tag{7.5}$$

and

$$(1 + v'^2)^{-1/2} = (1 + \phi_0'^2)^{-1/2} \left(1 + \sum_{\mu=1}^{\infty} \binom{-1/2}{\mu} Q^\mu \right).$$

Then

$$v'(1 + v'^2)^{-1/2} = (\phi_0' + \sum_{k=1}^{N} \phi_k' B_k)(1 + \phi_0'^2)^{-1/2} \left(1 + \sum_{\mu=1}^{\infty} \binom{-1/2}{\mu} Q^\mu \right)$$

$$= (1 + \phi_0'^2)^{-1/2} \left(\phi_0' + \sum_{k=1}^{N} \phi_k' B_k + \sum_{\mu=1}^{\infty} \binom{-1/2}{\mu} \phi_0' Q^\mu \right.$$

$$\left. + \sum_{k=1}^{N} \sum_{\mu=1}^{\infty} \binom{-1/2}{\mu} \phi_k' B^k Q^\mu \right).$$

For simplicity we consider here the case $N = 1$. For the general case $N \ge 1$ see [Miersemann (1994)]. Set

$$v = \frac{C}{B} + \phi_0 + \phi_1 B,$$

then

$$v'(1 + v'^2)^{-1/2} = (1 + \phi_0'^2)^{-1/2} \left(\phi_0' + \frac{\phi_1'}{1 + \phi_0'^2} B + O(B^2) \right),$$

and the boundary condition (7.4) implies that

$$\lim_{r \to 1-0} \frac{\phi_0'}{(1 + \phi_0'^2)^{1/2}} = \beta, \quad \phi_0'(0) = 0,$$

$$\lim_{r \to 1-0} \frac{\phi_1'}{(1 + \phi_0'^2)^{3/2}} = 0, \quad \phi_1'(0) = 0.$$

From

$$\operatorname{div} Tv - Bv = \frac{1}{r} \left(\frac{rv'}{\sqrt{1 + v'^2}} \right)' - Bv$$

$$= \frac{1}{r} \left(\frac{r\phi_0'}{\sqrt{1 + \phi_0'^2}} \right)' - C$$

$$+ B \left\{ \frac{1}{r} \left(r\phi_1'(1 + \phi_0'^2)^{-3/2} \right)' - \phi_0 \right\} + O(B^2)$$

and the requirement

$$\operatorname{div} Tv - Bv = O(B^2)$$

as $B \to 0$, we find the boundary value problems

$$\frac{1}{r} \left(\frac{r\phi_0'}{\sqrt{1 + \phi_0'^2}} \right)' = C \quad \text{on } (0, 1), \tag{7.6}$$

$$\lim_{r \to 1-0} \frac{\phi_0'}{\sqrt{1 + \phi_0'^2}} = \beta, \quad \phi_0'(0) = 0 \tag{7.7}$$

and

$$\frac{1}{r} \left(\frac{r\phi_1'}{(1 + \phi_0'^2)^{3/2}} \right)' = \phi_0 \quad \text{on } (0, 1), \tag{7.8}$$

$$\lim_{r \to 1-0} \frac{\phi_1'}{(1 + \phi_0'^2)^{3/2}} = 0, \quad \phi_1'(0) = 0. \tag{7.9}$$

From the boundary value problem (7.6), (7.7) we obtain

$$\phi_0(r) = \Phi_0(r) + K,$$

where

$$\Phi_0(r) = -\frac{1}{\beta} \sqrt{1 - \beta^2 r^2},$$

and K is a constant which is determined through the boundary value problem (7.8), (7.9) which defines the next term ϕ_1, up to a constant, in the asymptotic expansion. An easy calculation shows that

$$K = \frac{2}{3} \frac{1 - \sin^3 \beta}{\cos^3 \beta}.$$

The function ϕ_1 is defined by (7.8) and (7.9) up to an unknown constant. This constant is fixed by the boundary value problem for the next term ϕ_2. The quotient Q, defined by (7.5), may be written as

$$Q = \sum_{k=1}^{N} a_k(\gamma; r) B^k + O(B^{N+1}),$$

where the functions a_k and the remainder are uniformly bounded with respect to $r \in (0, 1)$ and $\gamma \in [0, \pi]$, see [Miersemann (1994)], p. 397. In particular, it follows that

$$\operatorname{div} T V_N - B V_N = O(B^{N+1}) \quad \text{in} \quad B_1(0),$$

$$\nu \cdot T V_N = \cos \gamma \quad \text{on} \quad \partial B_1(0),$$

where the remainder $O(B^{N+1})$ is uniformly bounded with respect to $r \in (0, 1)$ and $\gamma \in [0, \pi]$. $\qquad\square$

Proof of Proposition 7.2. The proof follows easily from the above lemma. We will show that for any given non-negative integer n the solution V of the boundary value problem (7.1), (7.2) satisfies

$$V = V_{n-1} + O(B^n)$$

uniformly with respect to $|\beta| \leq 1$ and $0 < r < 1$, where $V_{-1} = 2 \cos \gamma / B$. We recall that $\beta = \cos \gamma$. Set

$$V^+ = V_n + A B^n,$$

where A is a positive constant which will be determined later. Then

$$\operatorname{div} T V^+ - B V^+ = \operatorname{div} T V_n - B V_n - A B^{n+1}$$

$$= O(B^{n+1}) - A B^{n+1}$$

$$\leq 0,$$

provided A is large enough. Then, since

$$\lim_{r \to 1-0} \frac{(V^+)'}{\sqrt{1 + (V^+)'^2}} = \beta,$$

it follows from the comparison principle Theorem 7.1 that $V \leq V^+$ on $(0, 1)$. The same reason yields a lower bound if $V^- := V_n - A B^n$. Thus

$$V = V_n + O(B^n)$$

$$= V_{n-1} + O(B^n).$$

$\qquad\square$

7.2.3 *Wide circular tube*

Here we consider the boundary value problem

$$\left(\frac{rw'}{\sqrt{1+w'^2}} \right)' = \kappa rw \quad 0 < r < R,$$

$$\lim_{r \to R-0} \frac{w'}{\sqrt{1+w'^2}} = \cos\gamma \quad w'(0) = 0,$$

where $0 \leq \gamma \leq \pi$ and the capillarity constant κ is positive.

In contrast to the above narrow tube we are here interested in the case that R is large.

A formal asymptotic solution of this boundary value problem for large $\sqrt{\kappa}R$ was calculated by [Concus (1968)] by using a boundary layer technique which goes back to [Laplace (1806)]. The idea is to assume that there is a central core region covering most of the base domain in which w is small, and a boundary layer region near the wall in which w' increases rapidly to its given boundary value. Matching the core and the boundary layer solutions in the transition circle determines the thickness of the boundary layer.

This method was used by [Perko (1973)] to prove that a certain boundary layer approximation is asymptotically correct. More precisely, it is shown that for each given boundary contact angle γ away from the critical angles $\gamma = 0$ or $\gamma = \pi$ the relative error in the ordinate and slope of this boundary-layer approximation is uniform of order $1/R \ \ln(1/R)$ as $R \to \infty$ for $0 < r < R$. A formal second order boundary layer approximation was calculated by [Rayleigh (1915)].

Away from the boundary, for example in $0 < r < R - (5/2) \ln R$, the slope and the ordinate decrease exponentially as $R \to \infty$, see [Siegel (1980)].

We will prove an explicit asymptotic formula when $\sqrt{\kappa}R$ tends to infinity. We get the expected result that the leading term defines the capillary surface over the half plane with the given boundary contact angle. More precisely, let $w(r; R, \gamma, \kappa)$ be the solution of the above boundary value problem and let $v(s; \gamma, \kappa)$ be the solution of the following boundary value problem

$$\left(\frac{v'}{\sqrt{1+v'^2}} \right)' = \kappa v \text{ on } 0 < s < \infty,$$

$$\lim_{s \to +0} \frac{v'}{\sqrt{1+v'^2}} = -\cos\gamma,$$

which defines the capillary surface over a half plane with the same boundary contact angles, see Sec. 3.3.1.

Proposition 7.3. *The inequality*

$$|w(r; R, \gamma, \kappa) - v(R - r; \gamma, \kappa)| \leq 2.1 \left| \frac{\pi}{2} - \gamma \right| \frac{1}{\kappa R}$$

holds uniformly in $r \in [0, R]$ *and* $\gamma \in [0, \pi]$, *provided that* $\sqrt{\kappa} R \geq 6.4$. *In particular,*

$$\left| w_1 - \sqrt{\frac{2}{\kappa}} \sqrt{1 - \sin \gamma} \right| \leq 2.1 \left| \frac{\pi}{2} - \gamma \right| \frac{1}{\kappa R},$$

where $w_1 = w(R; R, \gamma, \kappa)$ *is the ascent of the capillary surface at the boundary of the tube.*

Proof. The proof is based on the comparison principle and on a mapping which brings the right parameter to the right place of the equation, see [Miersemann (1997)]. Instead of the above boundary value problem, we consider the normalized problem

$$\left(\frac{\rho u'(\rho)}{\sqrt{1 + u'(\rho)^2}} \right)' = \rho u(\rho), \quad 0 < \rho < M,$$

$$\lim_{\rho \to M - 0} \frac{u'(\rho)}{\sqrt{1 + u'(\rho)^2}} = \cos \gamma, \quad u'(0) = 0,$$

where

$$M = \sqrt{\kappa} R, \quad \rho = \sqrt{\kappa} r, \quad u(\rho) = \sqrt{\kappa} w(r).$$

Then $v(s) = u(M - s; M, \gamma)$ solves the boundary value problem

$$\frac{1}{M - s} \left(\frac{(M - s) v'(s)}{\sqrt{1 + v'(s)^2}} \right)' = v(s), \quad 0 < s < M,$$

$$\lim_{s \to +0} \frac{v'(s)}{\sqrt{1 + v'(s)^2}} = -\cos \gamma, \quad v'(M) = 0.$$

This boundary value problem becomes singular if $s = M$. Then we find an approximate solution $v_n(s; M, \gamma)$ in powers of $1/M^k$, $k \geq 0$, such that

$$U_n(x) := v_n(M - |x|; M, \gamma)$$

satisfies, see [Miersemann (1997)],

$$|\text{div } TU_n - U_n| \leq \frac{c_{n+1}}{M^{n+1}} \quad \text{in } B_M(0),$$

$$\lim_{|x| \to M - 0} \nu \cdot TU_n = \cos \gamma \quad \text{on } \partial B_M(0).$$

Using the comparison function

$$U^+ = U_n + \frac{c_{n+1}}{M^{n+1}},$$

we obtain

$$\operatorname{div} TU^+ - U^+ \le 0 \ \text{ in } \ B_M(0),$$
$$\lim_{|x| \to M-0} \nu \cdot TU^+ = \cos\gamma \ \text{ on } \ \partial B_M(0).$$

Since the function $U(x) = u(|x|; M, \gamma)$ satisfies

$$\operatorname{div} TU - U = 0 \ \text{ in } \ B_M(0),$$
$$\lim_{|x| \to M-0} \nu \cdot TU = \cos\gamma \ \text{ on } \ \partial B_M(0),$$

the comparison principle (Theorem 7.1) implies that

$$U(x) \le U^+(x) \ \text{ in } \ B_M(0) \ .$$

Analogously, the comparison function

$$U^- = U_n - \frac{c_{n+1}}{M^{n+1}}$$

yields a lower bound of $U(x)$ in $B_M(0)$. □

Remark 7.7. It can be shown, see [Miersemann (1997)], that there exists a complete asymptotic expansion in powers of $1/M^k$, $k \ge 0$. More precisely, there is an approximate solution of the type

$$v_n(s; M, \gamma) = \sum_{k=0}^{n} \phi_k(s; M, \gamma) \frac{1}{M^k},$$

where ψ_k, $k \ge 1$, are defined through linear boundary value problems, see [Miersemann (1997)], and for a given integer $n \ge 0$ there exists a constant c_{n+1} such that

$$|u(\rho; M, \gamma) - v_n(M - \rho; M, \gamma)| \le \frac{c_{n+1}}{M^{n+1}}$$

uniformly in $\rho \in [0, M]$ and $\gamma \in [0, \pi]$ where $u(\rho; M, \gamma)$ is defined in the previous sketch of the proof.

7.2.4   *Ascent at a needle*

We seek a surface S: $U = U(x)$, $x = (x_1, x_2)$, defined over the base domain $\Omega := \mathbb{R}^2 \setminus \overline{B_a(0)}$, where $B_a(0)$ is a disk with small radius a and the center at $x = 0$, such that U satisfies the nonlinear elliptic boundary value problem.

$$\operatorname{div} TU = \kappa\, U \qquad \text{in} \quad \Omega\,,$$
$$\nu \cdot TU = \cos\theta \qquad \text{on} \quad \partial\Omega\,,$$

where

$$TU = \frac{\nabla U}{\sqrt{1 + |\nabla U|^2}}\,,$$

$\kappa = \rho g/\sigma$ (ρ = density change across free surface, g = gravitational acceleration, σ = surface tension) is the (positive) capillarity constant. We assume that the gravity is positive and directed downwards. Further, θ in $0 \le \theta \le \pi$ denotes in this subsection the constant contact angle between the capillary surface and the cylinder with cross section $B_a(0)$. The vector ν is the exterior unit normal on $\partial\Omega$, that is, the interior normal on $\partial B_a(0)$. No explicit solution of the above boundary value problem is known. It was shown by [Johnson and Perko (1968)] that there exists a radially symmetric solution. From a maximum principle of Finn and Hwang (Theorem 7.2) for unbounded domains it follows that this symmetric solution is the only one.

Set

$$u(r) = U(x), \; r = \sqrt{x_1^2 + x_2^2}.$$

Then we have, see [Miersemann (2006)],

Proposition 7.4. *Set $B = \kappa a^2$ and let $\gamma = 0.5772\dots$ be the Euler constant. Then the ascent $u(a)$ of a liquid at a circular needle with radius a satisfies*

$$\frac{u(a)}{a} = -\cos\theta\;\left(\frac{1}{2}\ln B + \gamma - 2\ln 2 + \ln(1 + \sin\theta) + O(B^{\frac{1}{5}}\ln^2 B)\right)$$

as $B \to 0$, uniformly in $\theta \in [0, \pi]$.

Uniformly means that the remainder satisfies $|O(B^{\frac{1}{5}}\ln^2 B)| \le cB^{\frac{1}{5}}|\ln^2 B|$ for all $0 < B \le B_0$, B_0 sufficiently small, where the constant c depends only on B_0 and not on the contact angle θ.

It is noteworthy that the special non-linearity of the problem implies that the expansion is uniform with respect to $\theta \in [0, \pi]$ although $|Du|$ tends to infinity as $\theta \to 0$ or $\theta \to \pi$ and therefore the differential equation will be singular on $\partial\Omega$. Moreover, we do not need any growth assumption at infinity, which is a further consequence of the strong non-linearity of the problem.

In the case of complete wetting, i. e., if $\theta = 0$, the formula

$$u(a) \sim -a \left(\frac{1}{2} \ln B - 0.809 \ldots \right)$$

as $a \to 0$ has been derived formally in [Derjaguin (1946)] by matching of expansions. We recall that $B = \kappa a^2$. Higher order approximations where obtained formally by [James (1974)] and [Lo (1983)], also by matching arguments.

A mathematical proof of the formula

$$u(a) \sim -\frac{1}{2} \cos\theta \, a \ln B$$

as $a \to 0$ was given in [Turkington (1980)].

Sketch of the proof of Proposition 7.4. Since the solution is radially symmetric, we have the boundary value problem

$$\frac{1}{r} \left(\frac{r u'(r)}{\sqrt{1 + (u'(r))^2}} \right)' = \kappa u(r) \quad \text{in} \quad a < r < \infty,$$

$$\lim_{r \to a+0} \frac{u'(r)}{\sqrt{1 + (u'(r))^2}} = -\cos\theta .$$

Set

$$r = as, \quad v(s) = \frac{1}{a} u(as), \quad B = \kappa a^2,$$

then the above problem is changed to a problem where the unknown function is defined on a fixed interval:

$$\frac{1}{s} \left(\frac{s v'(s)}{\sqrt{1 + (v'(s))^2}} \right)' = B v(s) \quad \text{in} \quad 1 < s < \infty ,$$

$$\lim_{s \to 1+0} \frac{v'(s)}{\sqrt{1 + (v'(s))^2}} = -\cos\theta .$$

Then we can find an upper and a lower C^1-solution of this boundary value problem. We obtain the lower and the upper solution by gluing together a boundary layer expansion near the boundary $s = 1$ with a second expansion which is valid far from the boundary such that the resulting function is in C^1. $\qquad\square$

Remark 7.8. This method of composing of functions defined on different annular domains was used in [Miersemann (1996)], where a numerical method for the circular tube was proposed, see Sec. 7.2.7. Using this numerical method we will get more precise results for the ascent at the needle or at the exterior of a circular tube. The above asymptotic formulas are rather rough.

7.2.5 *Narrow tube of general bounded cross section*

Here we consider capillary tubes with general cross section which is constant along the cylinder. In contrast to the case of circular cross section the situation is more delicate. Let $\Omega \subset \mathbb{R}^2$ be a bounded domain with a piecewise $C^{2,\lambda}$ boundary and let u be a solution of the boundary value problem

$$\operatorname{div} Tu = Bu \quad \text{in } \Omega, \tag{7.10}$$

$$\nu \cdot Tu = \cos\gamma \tag{7.11}$$

at the smooth parts of $\partial\Omega$. Here is $Tu = \nabla u/\sqrt{1+|\nabla u|^2}$, B a positive constant (Bond number), and γ, $0 \le \gamma < \pi/2$, the constant contact angle between the capillary surface and the container wall. The vector ν denotes the exterior unit normal on the smooth parts of $\partial\Omega$. A formal ansatz

$$u = \frac{C}{B} + w_0(x,\gamma) + w_1(x,\gamma)B + \dots$$

in powers of small B, where C is a constant, leads to

$$\operatorname{div} Tw_0 = C \quad \text{in } \Omega, \tag{7.12}$$

$$\nu \cdot Tw_0 = \cos\gamma \tag{7.13}$$

at the smooth parts of $\partial\Omega$. Suppose a sufficiently regular w_0 is a solution of the boundary value problem (7.12) and (7.13), then we find after integration of equation (7.12) over Ω that

$$C = \frac{|\partial\Omega|}{|\Omega|}\cos\gamma. \tag{7.14}$$

A solution of the boundary value problem (7.12) and (7.13), where C satisfies (7.14), is said to be a *zero gravity solution*. Under the assumption that $\partial\Omega \in C^{2,\lambda}$ and that there is a zero gravity solution, [Siegel (1987)] has shown that

$$u(x,\gamma,B) = \frac{|\partial\Omega|\cos\gamma}{|\Omega|B} + w_0(x,\gamma) + O(B)$$

for each fixed γ, $0 < \gamma < \pi/2$, as B tends to zero. Here $w_0(x, \gamma)$ denotes the zero gravity solution which satisfies the side condition

$$\int_\Omega w_0(x, \gamma) \, dx = 0.$$

In the following we suppose that Ω is a bounded domain with a $C^{2,\lambda}$ boundary or with a boundary consisting of a finite number of $C^{2,\lambda}$ curves $\Sigma_1, \ldots, \Sigma_M$ and that Σ_l and Σ_{l+1} meet at an interior angle $2\alpha_l$ such that

$$0 < \alpha_l < \pi/2, \quad \alpha_l + \gamma > \pi/2.$$

Set $k_l = \sin \alpha_l / \cos \gamma$, define ω_l, $0 < \omega_l < \pi/2$, by

$$\tan \omega_l = \sqrt{1 - \tan \alpha_l / k_l^2},$$

and suppose that $2\omega_l < \pi/2$. Then we get, see [Miersemann (1993)],

Proposition 7.5. *Suppose that there exists a zero gravity $C^{2,\lambda}(\overline{\Omega})$ solution. Then there are $n + 1$ functions $w_l \in C^{2,\lambda}(\overline{\Omega})$, $l = 0, \ldots, n$, analytical in Ω and depending on γ, such that for fixed γ, $0 < \gamma < \pi/2$,*

$$u(x, \gamma, B) = \frac{|\partial \Omega| \cos \gamma}{|\Omega| B} + w_0(x, \gamma) + \sum_{l=1}^{n} w_l(x, \gamma) B^l + O(B^{n+1})$$

uniformly in $\overline{\Omega}$ as B tends to zero.

Remark 7.9. In the case that there is no classical zero gravity solution the asymptotic behavior as $B \to 0$ is more complicated, see [Finn (1986)], Chap. 6, [Tam (1986)] and [Scholz (2004)].

In the following two examples there exists a zero gravity solution which can be written in terms of elementary functions, elliptic integrals or elliptic functions

Example 7.1 (Annulus). Set $\Omega = \{x \in \mathbb{R}^2 : q < |x| < 1\}$ for a fixed q, $0 < q < 1$. From the comparison principle Theorem 7.1 it follows that each solution is radially symmetric and uniquely determined. A radially solution exists, see [Johnson and Perko (1968)]. Set $r = |x|$, then

$$w_0^*(r) = c \int_q^r \frac{s^2 - q}{\sqrt{s^2 - c^2(s^2 - q)^2}} \, ds,$$

where $c = \cos \gamma / (1 - q)$, is a zero gravity solution. This integral can be expressed in terms of elliptic integrals, see [Byrd and Friedman (1971)], formulas 218.00 and 218.01. The asymptotic expansion is here

$$u(x, \gamma, B) = \frac{2 \cos \gamma}{(1 - q) B} + \sum_{l=0}^{n} w_l(x, \gamma) B^l + O(B^{n+1})$$

as $B \to 0$, where $w_l(x, \gamma)$ are radially symmetric functions and w_0 is given by $w_0(x, \gamma) = w_0^*(r) + C_0$, where

$$C_0 = -\frac{2}{1-q^2} \int_q^1 r w_0^*(r) \, dr.$$

This integral can be expressed by elliptic integrals and Jacobian elliptic functions, see [Byrd and Friedman (1971)], formulas 218.06 and 314.04.

Example 7.2 (Regular n-gon). Let Ω be a regular n-gon, where the corners are located on the unit circle. It is easily seen that the lower hemisphere defined by

$$w_0^*(x, \gamma) = -\frac{1}{H}\sqrt{1 - H^2|x|^2},$$

where $H = \cos\gamma / \cos(\pi/n)$, is a solution of the zero gravity problem, which is analytical on $\overline{\Omega}$, provided that $H < 1$, or equivalently, that $\alpha_l + \gamma > \pi/2$ is satisfied. The assumption $2\omega_l < \pi/2$ holds in the cases $n = 3$ and $n = 4$ for each γ, $0 < \gamma < \pi/2$. If $n \geq 5$, then $2\omega_l < \pi/2$ is a condition on the boundary contact angle γ. Then, according to Proposition 7.5, we get the asymptotic formula

$$u(x, \gamma, B) = \frac{2\cos\gamma}{B\cos(\pi/n)} + \sum_{l=0}^{n} w_l(x, \gamma)B^l + O(B^{n+1}),$$

where $w_0(x, \gamma) = w_0^*(x, \gamma) + C_0$, with a constant C_0 such that

$$\int_\Omega w_0(x, \gamma) \, dx = 0.$$

After some calculation we find that

$$C_0 = \frac{2\pi}{3n\sin(\pi/n)\cos^3\gamma}\left(1 - \frac{n}{\pi}\int_0^{\pi/n}\left(1 - \frac{\cos^2\gamma}{\cos^2\theta}\right)^{3/2} d\theta\right).$$

This integral can be expressed in terms of elementary functions.

7.2.6 Ascent in a wedge

Brooke Taylor [Taylor (1712)] conjectured that the trace of the capillary interface on the walls of a narrow wedge is a hyperbola, see Fig. 7.5.

Here we consider the boundary value problem

$$\text{div } Tu = \kappa u \quad \text{in } \Omega, \tag{7.15}$$

$$\nu \cdot Tu = \cos\gamma \tag{7.16}$$

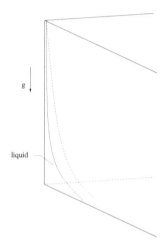

Fig. 7.5   Trace of the capillary surface

on the smooth parts of $\partial\Omega$, where $\kappa = const. > 0$ and ν is the exterior unit normal on the smooth parts of $\partial\Omega$. Let $x = 0$ be a corner of Ω with the interior angle 2α satisfying $0 < 2\alpha < \pi$. For simplicity we assume that the corner is bounded by lines near $x = 0$ and that $\Omega_R = \Omega \cap B_R$ coincides with the circular sector

$$\{x \in \mathbb{R}^2 : x_1 > 0, \ -x_1 \tan \alpha \le x_2 \le x_1 \tan \alpha\} \cap B_R$$

for a sufficiently small R, where B_R denotes the disk with center at the origin and radius R, see Fig. 7.6.

Furthermore, we assume that the contact angle satisfies $0 \le \gamma < \pi/2$. It was shown by [Concus and Finn (1970, 1974)] that every solution of (7.15) and (7.16) is bounded near the corner if and only if $\alpha + \gamma \ge \pi/2$ is satisfied and that in the unbounded case $\alpha + \gamma < \pi/2$

$$u(x; \alpha, \gamma) = \frac{h_{-1}(\theta; \alpha, \gamma)}{r} + O(1)$$

as $r \to 0$, where r, θ are polar coordinates centered at $x = 0$, and h_{-1} is defined by

$$h_{-1}(\theta; \alpha, \gamma) = \frac{\cos \theta - \sqrt{k^2 - \sin^2 \theta}}{\kappa k}$$

with $k = \sin \alpha / \cos \gamma$. The previous asymptotic formula was improved in [Miersemann (1989)] by showing that

$$u(x; \alpha, \gamma) = \frac{h_{-1}(\theta; \alpha, \gamma)}{r} + O(r^\epsilon)$$

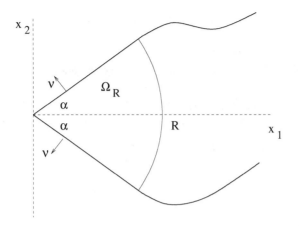

Fig. 7.6   Corner domain

for an $\epsilon > 0$. In fact, there exists an asymptotic expansion of u in powers of r^{4l-1}, $l = 0, 1, \ldots$, see [Miersemann (1993)]. In particular

$$u(x; \alpha, \gamma) = \frac{h_{-1}(\theta; \alpha, \gamma)}{r} + O(r^3)$$

as $r \to 0$.

Proposition 7.6. *For a given non-negative integer m there exist positive constants r_0, A and $m + 1$ functions $h_{4l-1}(\theta; \alpha, \gamma)$, $l = 0, \ldots, m$, analytical on (α, α) and bounded on $[\alpha, \alpha]$, such that*

$$\left| u(x) - \sum_{l=0}^{m} h_{4l-1}(\theta; \alpha, \gamma) r^{4l-1} \right| \le A \, r^{4m+3}$$

in Ω_{r_0}. Moreover, the constants r_0, A and the functions $h_{4l-1}(\theta; \alpha, \gamma)$ do not depend on the solution considered.

Proof. We assume that $\gamma > 0$. The proof is by induction. A function, we omit the arguments γ and α in the following,

$$v_n(x) := \sum_{l=0}^{n} h_{4l-1}(\theta) r^{4l-1}$$

is said to be an *approximate solution of order n* if

$$\operatorname{div} T v_n - \kappa v_n = O(r^{4n+3}) \quad \text{in} \quad \Omega_R,$$

$$\nu \cdot T v_n - \cos \gamma = O(r^{4n+3}) \quad \text{on} \quad \Sigma_R,$$

as $r \to 0$, where $\Sigma_R = (\partial \Omega \cap \overline{\Omega}) \setminus \{0\}$.

(i) $v_0 := h_{-1}(\theta)r^{-1}$ is an approximate solution of order $n = -1$.

This follows from a calculation and by exploiting the strong non-linearity of the problem. In contrast to this estimate one gets $\triangle v_0 \sim r^{-3}$ as $r \to 0$.

(ii) *Assume that v_n is an approximate solution, then there exist positive constants A, r_0, independent on the solution considered, such that*

$$|u(x) - v_n(x)| \le A\, r^{4n+3}$$

in Ω_{r_0}.

For a proof set

$$w = v + Aq(\theta)r^\lambda,$$

where $v = v_n$, $A = $ const., $\lambda = $ const. > 0 and $q \in C^2[-\alpha, \alpha]$. The positive constant λ and the function $q(\theta)$ will be determined later on independent of the constant A. In polar coordinates one finds that

$$\operatorname{div} Tw = r^{-1}\left\{ \left(\frac{rw_r}{\sqrt{1+|Dw|^2}}\right)_r + \left(\frac{r^{-1}w_\theta}{\sqrt{1+|Dw|^2}}\right)_\theta \right\},$$

where $|Dw| = w_r^2 + r^{-2}w_\theta^2$. The definition of w yields

$$\operatorname{div} Tw = r^{-1}\left\{ \left[\frac{rv_r + A\lambda qr^\lambda}{\sqrt{1+|Dw|^2}}\left(1 + \frac{Q}{1+|Dv|^2}\right)^{1/2}\right]_r \right.$$
$$\left. + \left[\frac{r^{-1}v_\theta + Aq'r^{\lambda-1}}{\sqrt{1+|Dw|^2}}\left(1 + \frac{Q}{1+|Dv|^2}\right)^{1/2}\right]_\theta \right\}$$

where

$$Q = 2A\lambda v_r qr^{\lambda-1} + 2Aq'v_\theta + A^2\lambda^2 q^2 r^{2\lambda-2} + A^2 q'^2 r^{2\lambda-2}.$$

Let for a given sufficiently small $\eta > 0$

$$|A|r_0^{\lambda+1} \le \eta.$$

After some calculations one finds that

$$\operatorname{div} Tw = \operatorname{div} Tv + AL_\lambda[q]r^\lambda + \eta_1 + \eta_2$$

in Ω_{r_0}. The linear differential operator L_λ is given by

$$L_\lambda[q] = (a_1(\theta)q' + a_0(\theta)q)' + b_1(\theta)q' + b_0(\theta)q,$$

where, with $h(\theta) := h_{-1}(\theta)$,

$$a_1 = h^2(h^2 + h'^2)^{-3/2}, \quad a_0 = \lambda h h'(h^2 + h'^2)^{-3/2},$$
$$b_1 = (\lambda + 2)h h'(h^2 + h'^2)^{-3/2}, \quad b_0 = \lambda(\lambda + 2)h'^2(h^2 + h'^2)^{-3/2},$$

and the η_1 and η_2 satisfy

$$|\eta_1| \leq c_1|A|r^{\lambda+4}, \quad |\eta_2| \leq c_2|A|^2 r^{2\lambda+1},$$

where the constants c_1 and c_2 do not depend on A or r. Since $v = v_n$ is an approximate solution of order n, we find that in Ω_{r_0}

$$\text{div } Tw - \kappa w = A(L_\lambda[q] - \kappa q)r^\lambda + \eta_1 + \eta_2 + \eta_3,$$

where

$$|\eta_3| \leq c_3 r^{4n+3},$$

c_3 is independent of A or r.

For $v \cdot Tw$ on Σ_{R_0} we have in polar coordinates

$$v \cdot Tw = \text{sign}(\theta)\frac{r^{-1}w_\theta}{\sqrt{1 + |Dw|^2}}, \quad \theta = \pm\alpha.$$

After some calculation we get

$$v \cdot Tw - \cos\gamma = AG_\lambda[q]r^{\lambda+1} + \mu_1 + \mu_2 + \mu_3$$

on Σ_{r_0}, where

$$G_\lambda[q] = a_1(\theta)q' + a_0(\theta)q$$

and

$$|\mu_1| \leq k_1|A|r^{\lambda+5}, \ |\mu_2| \leq k_2 A^2 r^{2\lambda+2}, \ |\mu_3| \leq k_3 r^{4n+4}$$

with constants k_j not dependent on A or r.

We insert $q = h(\theta)^{-\lambda} + \epsilon_0$ with $\lambda = 4n + 3$ into $L_\lambda[q]$ and $G_\lambda[q]$, where ϵ_0 is a sufficiently small positive constant such that $L_\lambda[q] - \kappa q < 0$ remains valid on $[-\alpha, \alpha]$.

If $A > 0$, then we find positive constants c_0 and k_0, not dependent on r or A, such that

$$\text{div } Tw - \kappa w \leq -c_0 A r^\lambda + c_1 A r^{\lambda+4} + c_2 A^2 r^{2\lambda+1} + c_3 r^\lambda$$

in Ω_{r_0} and

$$Tw - \cos\gamma \geq k_0 A r^{\lambda+1} - k_1 A r^{\lambda+5} - k_2 A^2 r^{2\lambda+2} - k_3 r^{\lambda+1}$$

on Σ_{r_0}. Since $|A|r_0^{\lambda+1} \leq \eta$, we find that

$$\text{div } Tw - \kappa w \leq -r^\lambda \left[A(c_0 - c_1 r r_0^4 - c_2\eta) - c_3\right]$$

in Ω_{r_0} and

$$Tw - \cos\gamma \geq r^{\lambda+1}\left[A(k_0 - k_1 r_0^4 - k_2\eta) - k_3\right]$$

on Σ_{r_0}. Let $A = A_0 r_0^{-\lambda}$ with a fixed sufficiently large positive constant which does not depend on r_0 or on the particular solution u considered. Then inequalities $|u - h_{-1}/r| \leq A$ or $|u - h_{-1}/r| \leq A r^\epsilon$ in Ω_{r_0} imply that

$$u \leq w \text{ on } \Gamma_{r_0}.$$

Finally we get from the above inequalities for sufficiently small $\eta > 0$ and $r_0 > 0$ that

$$\operatorname{div} Tw - \kappa w \leq 0 \quad \text{in } \Omega_{r_0} = \Omega \cap B_{r_0},$$
$$\nu \cdot Tw \geq \cos\gamma \quad \text{on } \Sigma_{r_0} = (\partial\Omega \cap B_{r_0}) \setminus \{0\},$$
$$w \geq u \quad \text{on } \Gamma_{r_0} = \Omega \cap \partial B_R.$$

The comparison principle of Concus and Finn (Theorem 7.1) implies that

$$u(x) \leq V_n(x) + C r^{4n+3},$$

where $C = A\left(h(\theta)^{-(4n+3)} + \epsilon_0\right)$. If $A < 0$, then we obtain by the same reasoning a lower bound for u in Ω_{r_0}. Thus the assertion (ii) is shown.

(iii) *Suppose that $v_m(x)$ is an approximate solution. Then there exist functions $h_{4m+3}(\theta)$, analytical in $(-\alpha, \alpha)$ and bounded on $[-\alpha, \alpha]$, such that*

$$v_{m+1}(x) = v_m(x) + h_{4m+3}(\theta) r^{4m+3}$$

is an approximate solution.

For a proof we set $v = v_m$ and $\lambda = 4m + 3$. We seek a function $q(\theta)$ such that $w = v_m + q(\theta) r^{4m+3}$ is an approximate solution of order $n = m + 1$. After some calculation we obtain

$$\operatorname{div} Tw - \kappa w = \operatorname{div} v - \kappa v + (L_\lambda[q] - \kappa q)r^\lambda + O\left(r^{\lambda+4}\right)$$

in Ω_{r_0}, and

$$\nu \cdot Tw = \nu \cdot Tv + G_\lambda[q]r^{\lambda+1} + O\left(r^{\lambda+5}\right)$$

on Σ_{r_0}, provided $r_0 > 0$ is sufficiently small. Since $v = v_m$ is an approximate solution, we get for an analytic function $f_\lambda(\theta)$ on $[-\alpha, \alpha]$ that

$$\operatorname{div} Tv - \kappa v = f_\lambda(\theta) r^\lambda + O\left(r^{\lambda+4}\right)$$

in Ω_{r_0}, and for constants $g_\lambda^{(1)}$ and $g_\lambda^{(2)}$ that

$$\nu \cdot Tv = \cos\gamma + g_\lambda^{(i)} r^{\lambda+1} + O\left(r^{\lambda+5}\right)$$

on $\Sigma_{r_0}^{(i)}$, where $\Sigma_{r_0}^{(1)}$ and $\Sigma_{r_0}^{(2)}$ denote the upper part resp. the lower part of Σ_{r_0}, see Fig. 7.6.

Thus, we seek a solution $q(\theta)$ of the linear and regular boundary value problem

$$L_\lambda[q] - \kappa q = f_\lambda(\theta) \text{ in } (-\alpha, \alpha),$$

$$G_\lambda[q] = \begin{cases} g_\lambda^{(1)} &, \quad \theta = \alpha, \\ g_\lambda^{(2)} &, \quad \theta = -\alpha, \end{cases}.$$

We will show that the associated homogeneous problem has only the zero solution. Consequently, there exists a unique solution of the inhomogeneous boundary problem.

Let $q_0(\theta)$ be a solution of the homogeneous problem. We replace $v = v_m$ by $v_m + q_0(\theta)r^\lambda \ln r$. Then we find that

$$\begin{aligned} \operatorname{div} Tw &= \operatorname{div} Tv + AL_\lambda[q]r^\lambda + \eta_1 + \eta_2 \\ &= \operatorname{div} Tv_m + L_\lambda[q_0]r^\lambda \ln r + AL_\lambda[q]r^\lambda + \eta_1 + \eta_2 + \eta_3. \end{aligned}$$

From the definition of w we get

$$\begin{aligned} \operatorname{div} Tw - \kappa w &= (L_\lambda[q_0] - \kappa q_0)\, r^\lambda \ln r \\ &\quad + A\,(L_\lambda[q] - \kappa q)\, r^\lambda + \eta_1 + \eta_2 + \eta_3. \end{aligned}$$

For the boundary value operator we have

$$\begin{aligned} \nu \cdot Tw - \cos\gamma &= G_\lambda[q_0]r^{\lambda+1}\ln r + AG_\lambda[q]r^{\lambda+1} \\ &\quad + \mu_1 + \mu_2 + \mu_3. \end{aligned}$$

By the same reasoning as above we get the asymptotic formula

$$u(x) = v_m(x) + q_0(\theta)r^\lambda \ln r + O(r^\lambda),$$

where $\lambda = 4m + 3$.

This expansion, together with the previous expansion

$$u(x) = v_m(x) + O(r^\lambda),$$

force that $q_0(\theta) = 0$ for all $\theta \in [-\alpha, \alpha]$. $\qquad\square$

Remark 7.10. For a proof in the case $\gamma = 0$ see [Miersemann (1993)], pp. 103–107. In fact, one can make the asymptotic expansion uniform with respect to $\gamma \to 0$, see an exercise to this chapter. In particular, the constant A in the estimate of Proposition 7.6 does not depend on γ, $0 < \gamma \leq \gamma_0$, where γ_0 satisfies the inequality $\alpha + \gamma_0 < \pi/2$.

The formula $u = h_{-1}(\theta; \alpha, \gamma)/r^{-1} + O(1)$ as $r \to 0$ implies a formula for the area of the capillary surface near the corner. Set $\Omega_{r,r_0} = \Omega_{r_0} \setminus \Omega_r$, $0 < r < r_0$.

Proposition 7.7. *Let $0 < 2\alpha < \pi$, $0 \le \gamma < \pi/2$ and $\alpha + \gamma < \pi/2$. Then*

$$\int_{\Omega_{r,r_0}} \sqrt{1 + |\nabla u|^2}\, dx = -\frac{2}{\kappa}\left(\frac{\pi}{2} - \alpha - \gamma\right) \ln r + O(1)$$

as $r \to 0$. The remainder $O(1)$ is independent on the solution considered and on γ, $0 < \gamma \le \gamma_0$, where $\alpha + \gamma_0 < \pi/2$.

Proof. From the differential equation we get

$$\int_{\Omega_{r,r_0}} u \operatorname{div} Tu\, dx = \kappa \int_{\Omega_{r,r_0}} u^2 dx.$$

Integration by parts implies

$$-\int_{\Omega_{r,r_0}} \frac{|\nabla u|^2}{\sqrt{1+|\nabla u|^2}} dx + \int_{\partial\Omega_{r,r_0}} u\, \nu \cdot Tu\, ds = \kappa \int_{\Omega_{r,r_0}} u^2 dx.$$

Then

$$\int_{\Omega_{r,r_0}} \sqrt{1+|\nabla u|^2}\, dx = J_1 + J_2 + J_3,$$

where

$$J_1 = \int_{\Omega_{r,r_0}} \frac{dx}{\sqrt{1+|\nabla u|^2}},$$

$$J_2 = \int_{\partial\Omega_{r,r_0}} u\, \nu \cdot Tu\, ds,$$

$$J_3 = -\kappa \int_{\Omega_{r,r_0}} u^2\, dx.$$

Set $h(\theta) := h_{-1}(\theta; \alpha, \gamma)$, then we find from the boundary condition and the asymptotic formula $u(x) = h(\theta)r^{-1} + O(1)$ as $r \to 0$ that

$$J_2 = 2h(\alpha)(\ln r_0 - \ln r)\cos\gamma = O(1).$$

The asymptotic formula implies

$$J_3 = -\kappa(\ln r - \ln r_0)\int_{-\alpha}^{\alpha} h^2(\theta)\, d\theta + O(1).$$

Summarizing, we obtain

$$\int_{\Omega_{r,r_0}} \sqrt{1+|\nabla u|^2}\, dx = -\left(2h(\alpha)\cos\gamma - \kappa\int_{-\alpha}^{\alpha} h^2(\theta)\, d\theta\right) \ln r + O(1).$$

After an easy calculation we arrive at the formula of the proposition. From an exercise it follows that the remainder is uniformly bounded if $\gamma \to 0$. $\quad\square$

Remark 7.11. Concerning the existence of a solution see [Emmer (1973)] when $\alpha + \gamma > \pi/2$ and [Finn and Gerhardt (1977)] if $\alpha + \gamma \leq \pi/2$. The case of different contact angles on the walls of the wedge was discussed in [Chen, Finn and Miersemann (2008)]. See also [Lancaster (2010)], where a conjecture of Concus and Finn was confirmed. This conjecture says that solutions are discontinuous near the corner if the contact angles γ_1, γ_2, $0 < \gamma_i < \pi$, satisfy $2\alpha < |\gamma_2 - \gamma_1| < \pi$.

7.2.7 *Ascent along a cusp*

Putting two cylinders together one gets a zero opening angle of the cross section, see Fig. 7.7. In the case of two circular cylinders with the same

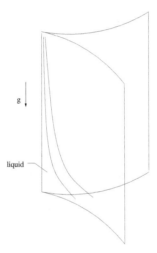

Fig. 7.7 Ascent along a cusp

cross section [Scholz (2003)] proved the asymptotic formula

$$u = \frac{C}{r^2} + O\left(\frac{1}{r}\right) \qquad \text{if } r \to 0,$$

where $C = C(R, \gamma)$ is a constant depending only the radius R of the cylinder and on the boundary contact angle γ.

7.2.8 *A numerical method*

This method is based on asymptotic expansions on subdomains of the domain in consideration.

7.2.8.1 *Rotationally symmetric tube*

Here we consider a rotationally symmetric capillary tube in the presence of gravity. By composing asymptotic expansions defined on different annular subdomains one obtains a numerical procedure for the calculation the capillary surface in a circular tube or between two coaxial circular tubes. Explicit error estimates uniform with respect to the boundary contact angle are given. The proof of this estimate is based on the comparison principle of Concus and Finn (Theorem 7.1).

Let Ω be the annular domain $a < \sqrt{x_1^2 + x_2^2} < b$, where $a \geq 0$. The solution is radially symmetric and satisfies the one-dimensional boundary value problem

$$\left(\frac{sw'(s)}{\sqrt{1 + w'(s)^2}}\right)' = \kappa sw(s) \quad \text{in} \quad a < s < b, \tag{7.17}$$

$$\lim_{s \to b - 0} \frac{w'(s)}{\sqrt{1 + w'(s)^2}} = \cos\gamma, \tag{7.18}$$

$$\lim_{r \to a + 0} \frac{w'(s)}{\sqrt{1 + w'(s)^2}} = -\cos\gamma^*, \tag{7.19}$$

where γ and γ^* are the given constant contact angles at the container walls. In the case of a capillary tube, i. e., if $a = 0$, then $\gamma^* = \pi/2$, or the boundary condition at a has to be replaced by $w'(0) = 0$.

Even in this radially symmetric case no explicit solution of the problem (7.17)–(7.19) is known except $w \equiv 0$ in the case $\gamma = \pi/2$.

Let $w(\kappa, s)$ be the solution of the above boundary value problem. Set $B = b^2\kappa$, $s = b\,r$, $r_0 = a/b$ and

$$w(\kappa, s) = b\,u(B, r),$$

then $u(B, r)$ is the solution of the problem

$$\left(\frac{ru'(r)}{\sqrt{1 + u'(r)^2}}\right)' = Bru(r) \quad \text{in} \quad r_0 < r < 1, \tag{7.20}$$

$$\lim_{r \to 1 - 0} \frac{u'(r)}{\sqrt{1 + u'(r)^2}} = \cos\gamma, \tag{7.21}$$

$$\lim_{r \to r_0 + 0} \frac{u'(r)}{\sqrt{1 + u'(r)^2}} = -\cos\gamma^*, \tag{7.22}$$

For an integer $n \geq 2$ let $0 \leq r_0 < r_1 < \ldots < r_{n-1} < r_n = 1$ be a subdivision of the interval $r_0 < r < 1$. On each subinterval $I_k : \quad r_{k-1} < r < r_k$,

$1 \leq k \leq n$, we consider the boundary value problem

$$\left(\frac{ru'(r)}{\sqrt{1 + u'(r)^2}} \right)' = Bru(r) \quad \text{in} \quad r_{k-1} < r < r_k, \tag{7.23}$$

$$\frac{u'(r)}{\sqrt{1 + u'(r)^2}} = \begin{cases} \cos \gamma_{k-1} & \text{if} \quad r = r_{k-1} \\ \cos \gamma_k & \text{if} \quad r = r_k \end{cases} \tag{7.24}$$

for angles $0 \leq \gamma_k \leq \pi$, $0 \leq k \leq n$, where $\gamma_0 = \pi - \gamma^*$, $\gamma_n = \gamma$, see Fig. 7.8. For given angles γ_{k-1}, γ_k there exists a unique solution of the

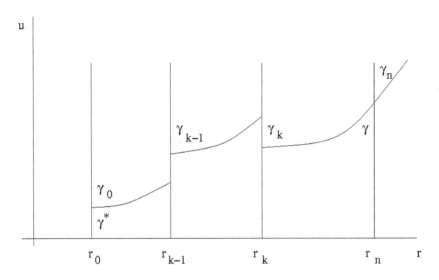

Fig. 7.8   Subdivision, notation

problem (7.23), (7.24). Concerning the question of existence of rotationally symmetric solutions see [Johnson and Perko (1968)]. Set $a_k = \cos \gamma_k$ and let

$$u_k(r) = u(B, r_{k-1}, r_k, a_{k-1}, a_k, r)$$

be the solution of the boundary value problem (7.23) and (7.24). By the same reasoning as in [Siegel (1980, 1987)] one obtains an asymptotic formula or even a complete asymptotic expansion of the solution $u_k(r)$ with respect to $h_k = r_k - r_{k-1}$ as $h_k \to 0$, see [Miersemann (1993, 1994)]. Moreover, for fixed r_{k-1}, r_k this expansion is uniformly with respect to the angle coordinates a_{k-1}, a_k from the closed interval $[-1, 1]$. This follows as in [Miersemann (1994)]. The finite asymptotic sum is given by

$$u_{(m)}(B, r_{k-1}, r_k, a_{k-1}, a_k, r) = \frac{C_{-1}(r_{k-1}, r_k, a_{k-1}, a_k)}{h_k B}$$

$$+ \sum_{l=0}^{m} \phi_l(r_{k-1}, r_k, a_{k-1}, a_k, \frac{r - r_{k-1}}{h_k}) B^l h_k^{2l+1},$$

where the constant C_{-1} and the functions ϕ_l are defined by a recurrent system of boundary value problems, see [Miersemann (1993)]. In particular, C_{-1} and ϕ_0 are given by the following formulas.

Set $v = r_{k-1}$, $w = r_k$, $x = a_{k-1}$ and $y = a_k$, then

$$C_{-1}(v, w, x, y) = -\frac{2v}{v+w}x + \frac{2w}{v+w}y. \tag{7.25}$$

Let

$$f(v, w, x, y, \zeta) = \frac{vx + (v\zeta + (w - v)\zeta^2/2)\,C_{-1}(v, w, x, y)}{v + (w - v)\zeta},$$

$$q_0(v, w, x, y, \zeta) = \frac{f(v, w, x, y, \zeta)}{\sqrt{1 - f^2(v, w, x, y, \zeta)}},$$

then

$$\phi_0(v, w, x, y, \zeta) = -\int_0^1 q_0(v, w, x, y, \tau)\left(1 - \frac{2v\tau + (w - v)\tau^2}{v + w}\right) d\tau$$

$$+ \int_0^\zeta q_0(v, w, x, y, \tau)\, d\tau\,. \tag{7.26}$$

This sum defines an approximate solution of the boundary value problem (7.23)–(7.24) in the sense that the sum satisfies the boundary condition (7.24) and solves the differential equation (7.23) up to a function which can be estimated by a positive power of Bh_k^2, see [Miersemann (1993, 1994)]. We define the composed function $U_{(m)}(B, \underline{r}, a, r)$ by

$$U_{(m)}(B, \underline{r}, a, r) = u_{(m)}(r_{k-1}, r_k, a_{k-1}, a_k, r),$$

where $r \in I_k$, $\underline{r} = (r_0, r_1, \ldots, r_{n-1}, r_n)$ and $a = (a_0, a_1, \ldots, a_{n-1}, a_n)$ with $a_0 = -\cos\gamma^*$, $a_n = \cos\gamma$.

This function $U_{(m)}(B, \underline{r}, a, r)$ is in $C^1[0, 1)$ if and only if the following system of $n - 1$ nonlinear equations, $k = 1, 2, \ldots, n - 1$, is satisfied.

$$\frac{C_{-1}(r_{k-1}, r_k, a_{k-1}, a_k)}{h_k B} + \sum_{j=0}^{m} \phi_j(r_{k-1}, r_k, a_{k-1}, a_k, 1) B^j h_k^{2j+1} \tag{7.27}$$

$$= \frac{C_{-1}(r_k, r_{k+1}, a_k, a_{k+1})}{h_{k+1} B} + \sum_{j=0}^{m} \phi_j(r_k, r_{k+1}, a_k, a_{k+1}, 0) B^j h_{k+1}^{2j+1}\,.$$

From now on we will consider the case $m = 0$, and that we have equidistant subdivisions $r_k = hk$, where $h = 1/n$. Under these assumptions there exists a solution of the system (7.27) if, roughly speaking, a smallness condition is satisfied for the product Bh. More precisely, set

$$\eta(h, \cos\gamma) = \frac{1}{\sqrt{1 - \cos\gamma} + \sqrt{1 - \cos\gamma + h(1 - h)\cos\gamma}}, \qquad (7.28)$$

then one concludes from formulas (7.25) and (7.26) which define C_{-1} and ϕ_0, that there exists a solution $0 \le a_1 \le a_2 \le \ldots \le a_{n-1} \le \cos\gamma$ of the system (7.27) if

$$6B\eta(h, \cos\gamma)h^2 \le 1$$

holds.

Set $Tu = \nabla u / \sqrt{1 + |\nabla u|^2}$. The function $\hat{U}_{(0)}(a, x) = U_{(0)}(B, \underline{r}, a, r)$, where $r = \sqrt{x_1^2 + r_2^2}$, satisfies on the union of the annular domains

$$\{r : (k - 1)h < r < kh\}, \quad k = 1, \ldots, n,$$

the inequality

$$\left| \text{div } T\hat{U}_{(0)} - B\hat{U}_{(0)} \right| \le c_1 Bh, \qquad (7.29)$$

where $c_1 = 4\eta(h, \cos\gamma)$. This estimate follows from the special properties of C_{-1} and ϕ_0, defined by (7.25) and (7.26). If there exists a solution of the nonlinear system (7.27) of the $n - 1$ equations then $\hat{U}_{(0,h,B)} \in C^1$ in the open annular domain Ω.

Under the assumption $\hat{U}_{(0)} \in C^1(B_1)$ one concludes by the same reasoning as in [Miersemann (1993, 1994)] from (7.29) and the comparison principle of Concus and Finn (Theorem 7.1) in its weak form an error estimate. We apply this comparison principle to the comparison functions $U^\pm = \hat{U}_{(0)} \pm c_1 h$. Since in Ω

$$\text{div } TU^+ - BU^+ \le 0$$

and

$$\lim_{|x| \to 1} \nu \cdot TU^+ = \cos\gamma$$

on $\partial\Omega$. Then $u \le U^+$ holds in Ω, where u denotes the solution of the nonlinear boundary value problem (7.20)–(7.21). By the same reasoning, the comparison function U^- yields a lower bound for w. Thus the above considerations are a sketch of the proof of the following proposition.

Proposition 7.8. *Suppose that $6B\eta(h, \cos\gamma)h^2 \le 1$, where η is defined by (7.28). Then there exists a solution $0 \le a_1 \le a_2 \le \ldots \le a_{n-1} \le \cos\gamma$ of the system (7.27), with $m = 0$ and $r_{k-1} - r_k = h = 1/n$, such that*

$$\max_{0 \le r \le 1} \left| U_{(0)}(a, r) - u(r) \right| \le 4\eta(h, \cos\gamma)h$$

holds.

Remark 7.12. The crucial point here is that this estimate yields an explicit error estimate of order $O(\sqrt{h})$ uniformly with respect to the boundary contact angle γ in $0 \le \gamma \le \pi/2$ despite the fact that u' becomes unbounded if $\gamma \to 0$ or $\gamma \to \pi$. The reason for this behavior is the strong non-linearity of the problem. There is no counterpart for linear problems.

Remark 7.13. It is worthy to note that already [Runge (1895)] proposed a scheme for the calculation of the solution of the boundary value problem (7.20)–(7.22). Another method was presented by [Concus and Pereyra (1983)]. Indeed, no error estimate uniform in γ is known.

7.2.8.2 *Newton scheme, rotational symmetric tube*

In the case of first order approximation, i. e., if $m = 0$, we solved numerically the nonlinear system (7.27) with Newton iteration by using standard software packages. Let $\underline{r} = (r_0, r_1, \ldots, r_n)$ with $a = (a_0, a_1, \ldots, a_{n-1}, a_n)$ with $a_0 = -\cos\gamma^*$ and $a_n = \cos\gamma$. Set $z = (a_1, \ldots, a_{n-1})$ and let $m = 0$ in the equation (7.27). Then the system (7.27) can be written as

$$Mz = Be(\underline{r}, a) + p,$$

where B is the Bond number, M the tridiagonal $(n-1, n-1)$–matrix with entries m_{kl}, $k, l = 1, 2, \ldots, n-1$, defined by

$$m_{ii} = \frac{2r_i(r_{i+1} - r_i)}{r_{i-1} + r_i} + \frac{2r_i(r_i - r_{i-1})}{r_i + r_{i+1}},$$

$$m_{i,i+1} = \frac{-2r_{i+1}(r_i - r_{i-1})}{r_i + r_{i+1}},$$

$$m_{i+1,i} = \frac{-2r_i(r_{i+2} - r_{i+1})}{r_i + r_{i+1}},$$

$$m_{kl} = 0 \quad \text{else}.$$

The $(n-1)$–vector e is given by the coordinates

$$e_k(\underline{r}, a) = (r_{k-1} - r_k)^2(r_k - r_{k-1})\phi_0(r_k, r_{k-1}, a_k, a_{k-1}, 0)$$
$$- (r_k - r_{k-1})^2(r_{k+1} - r_k)\phi_0(r_{k-1}, r_k, a_{k-1}, a_k, 1),$$

$k = 1, \ldots, n-1$. The function ϕ_0 is defined through formula (7.26) and can be calculated by using a standard numerical procedure.

The coordinates of the $(n-1)$–vector p are $p_{n-1} = 2\cos\gamma(r_{n-1} - r_{n-2})/(r_{n-1} + 1)$ and $p_k = 0$ if $k = 1, \ldots, n-2$. The associated Newton iteration scheme is defined by

$$\left(M - BDe(\underline{r}, a^{(n)})\right) z^{(n+1)} = p + B\left(e(\underline{r}, a^{(n)}) - De(\underline{r}, a^{(n)}) z^{(n)}\right)$$

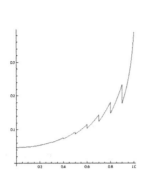 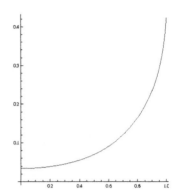

Fig. 7.9 Composed functions after one and two Newton iterations

with an initial vector $a^{(0)}$ given by $a_k^{(0)} = (k/n)\cos\gamma$, $k = 0, 1, \ldots, n$, for example. Here De denotes the Jacobian matrix of $e(\underline{r}, a)$ with the entries $\partial e_k/\partial a_l$, $k, l = 1, \ldots, n-1$, De is a tridiagonal matrix. Replacing the matrix $De(\underline{r}, a^{(n)})$ by $De(\underline{r}, a^{(0)})$, one obtains a simplified Newton scheme which runs faster.

Fig. 7.9 show composed functions in the case of a capillary tube with 10 equidistant subdivisions of the interval $(0, 1)$, $B = 13.4$ and $\cos\gamma = 0.99$, after one and two iterations, resp.

7.2.8.3 *Capillary surface between two parallel plates*

Following the above considerations for the circular tube, we get a numerical method which runs very well. The capillary surface between two vertical parallel plates of distance d is defined through the one-dimensional boundary value problem

$$\left(\frac{w'(s)}{\sqrt{1 + w'(s)^2}}\right)' = \kappa w(s) \quad \text{in} \quad 0 < s < d, \tag{7.30}$$

$$\lim_{s \to d-0} \frac{w'(s)}{\sqrt{1 + w'(s)^2}} = \cos\gamma, \tag{7.31}$$

$$\lim_{r \to +0} \frac{w'(s)}{\sqrt{1 + w'(s)^2}} = -\cos\gamma^*, \tag{7.32}$$

where γ and γ^* are the given constant contact angles at the plates.

Let $w(\kappa, s)$ be the solution of the above boundary value problem. Set $B = d^2\kappa$, $s = d\,r$ and

$$w(\kappa, s) = d\,u(B, r).$$

Then $u(B, r)$ is the solution of the problem

$$\left(\frac{u'(r)}{\sqrt{1 + u'(r)^2}}\right)' = Bu(r) \quad \text{in} \quad 0 < r < 1, \tag{7.33}$$

$$\lim_{r \to 1-0} \frac{u'(r)}{\sqrt{1 + u'(r)^2}} = \cos \gamma, \tag{7.34}$$

$$\lim_{r \to +0} \frac{u'(r)}{\sqrt{1 + u'(r)^2}} = -\cos \gamma^*, \tag{7.35}$$

where γ and γ^* are the given constant contact angles at the plates, see Fig. 7.8 with $r_0 = 0$.

For an integer $n \geq 2$ let $0 = r_0 < r_1 < \dots < r_{n-1} < r_n = 1$ be a subdivision of the interval $r_0 < r < 1$. On each subinterval I_k : $r_{k-1} < r < r_k$, $1 \leq k \leq n$, we consider the boundary value problem

$$\left(\frac{u'(r)}{\sqrt{1 + u'(r)^2}}\right)' = Bu(r) \quad \text{in} \quad r_{k-1} < r < r_k, \tag{7.36}$$

$$\frac{u'(r)}{\sqrt{1 + u'(r)^2}} = \begin{cases} \cos \gamma_{k-1} & \text{if} \quad r = r_{k-1} \\ \cos \gamma_k & \text{if} \quad r = r_k \end{cases} \tag{7.37}$$

for angles $0 \leq \gamma_k \leq \pi$, $0 \leq k \leq n$, where $\gamma_0 = \pi - \gamma^*$, $\gamma_n = \gamma$, see Fig. 7.8 with $r_0 = 0$. Set $a_k = \cos \gamma_k$ and let

$$u_k(r) = u(B, r_{k-1}, r_k, a_{k-1}, a_k, r)$$

be the solution of (7.36) and (7.37). The first two terms of the asymptotic expansion are given by

$$\frac{C_{-1}(a_{k-1}, a_k)}{h_k B} + \phi_0(a_{k-1}, a_k, \frac{r - r_{k-1}}{h_k}) h_k,$$

where

$$C_{-1}(x, y) = y - x,$$

$$\phi_0(x, y, \zeta) = \int_0^\zeta q_0(x, y, \tau) \, d\tau - \int_0^1 (1 - \tau) q_0(x, y, \tau) \, d\tau. \tag{7.38}$$

The function q_0 is defined through

$$q_0(x, y, \zeta) = \frac{f(x, y, \zeta)}{\sqrt{1 - f^2(x, y, \zeta)}},$$

where

$$f(x, y, \zeta) = (y - x)\zeta + x.$$

The same reasoning as in the case of the circular tube yields a system of $n-1$ nonlinear equations defined by

$$\frac{C_{-1}(a_{k-1},a_k)}{h_k B} + \phi_0(a_{k-1},a_k,1)h_k \tag{7.39}$$

$$= \frac{C_{-1}(a_k,a_{k+1})}{h_{k+1}B} + \phi_0(a_k,a_{k+1},0)h_{k+1},$$

where $k=1,2,\ldots,n-1$. The function ϕ_0 is given through formula (7.38) and can be evaluated by numerical integration.

7.2.8.4 *Newton scheme, parallel plates*

Let $\underline{r}=(r_0,r_1,\ldots,r_n)$, $a=(a_0,a_1,\ldots,a_{n-1},a_n)$ with $a_0=-\cos\gamma^*$ and $a_n=\cos\gamma$. Set $z=(a_1,\ldots,a_{n-1})$. Then the system (7.39) can be written as

$$Mz = Be(\underline{r},a)+p,$$

where B is the Bond number, M the tridiagonal $(n-1,n-1)$–matrix with entries m_{kl}, $k,l=1,2,\ldots,n-1$, defined by

$$m_{ii} = r_{i+1}-r_{i-1},$$
$$m_{i,i+1} = -(r_i-r_{i-1}),$$
$$m_{i+1,i} = -(r_{i+2}-r_{i+1}),$$
$$m_{kl} = 0 \quad \text{else}.$$

The $(n-1)$–vector e is given by the coordinates

$$e_k(\underline{r},a) = (r_{k-1}-r_k)^2(r_k-r_{k-1})\phi_0(a_k,a_{k-1},0)$$
$$-(r_k-r_{k-1})^2(r_{k+1}-r_k)\phi_0(a_{k-1},a_k,1),$$

$k=1,\ldots,n-1$. The coordinates of the $(n-1)$–vector p are $p_1=a_0(r_2-r_1)$, $p_{n-1}=a_n(r_{n-1}-r_{n-2})$ and $p_k=0$ if $k=2,\ldots,n-2$. The associated Newton iteration scheme is defined by

$$\left(M - BDe(\underline{r},a^{(n)})\right)z^{(n+1)} = p+B\left(e(\underline{r},a^{(n)})-De(\underline{r},a^{(n)})z^{(n)}\right)$$

with an initial vector $a^{(0)}$ given by $a_k^{(0)}=(k/n)\cos\gamma$, $k=0,1,\ldots,n$, for example. Here De denotes the Jacobian matrix of $e(\underline{r},a)$ with the entries $\partial e_k/\partial a_l$, $k,l=1,\ldots,n-1$, and De is a tridiagonal matrix. Replacing the matrix $De(\underline{r},a^{(n)})$ by $De(\underline{r},a^{(0)})$, one obtains a simplified Newton scheme which runs faster.

7.2.8.5 *General cross section*

The main idea of the method can be extended to the capillary tube problem with general cross section. An open problem is here to find error estimates. In the presence of gravity, the equilibrium free surface $S : u = u(x_1, x_2)$ of a liquid inside a tube satisfies the nonlinear elliptic Neumann problem

$$\operatorname{div} \frac{\nabla u}{\sqrt{1 + |\nabla u|^2}} = Bu \quad \text{in} \quad \Omega, \tag{7.40}$$

$$\nu \cdot \frac{\nabla u}{\sqrt{1 + |\nabla u|^2}} = \cos \gamma \quad \text{on} \quad \partial\Omega,$$

where $\Omega \subset \mathbb{R}^2$ is the cross section of the tube, B is the Bond number which is a positive constant, and γ in $0 \le \gamma \le \pi$ denotes the constant contact angle between the capillary surface and the cylinder wall. The vector ν is the exterior unit normal at $\partial\Omega$.

Thus, the problem of finding a capillary surface is a geometric one: to find a surface whose mean curvature is a prescribed function of position and which meets prescribed boundary walls in a prescribed angle γ, see a remark in [Finn (1986)], p. 12. We recall that div $Tu \equiv 2H(x)$, where $H(x)$ is the mean curvature of the surface S: $z = u(x)$ at $(x, u(x))$.

This observation leads to the following scheme for the numerical calculation of the capillary surface. Consider a triangulation of a two dimensional domain Ω with, for simplicity, a polygonal boundary. We assume that adjoining triangles join either a single corner or an entire side. Let $z_k(x)$ be a hemisphere over a ball B_{r_k} with radius r_k. By ν_k we denote the outer unit normal at the boundary $\partial\Omega_k \backslash \{\text{corners}\}$ of the triangle Ω_k. Suppose that

(i) $\Omega_k \subseteq B_{r_k}$ for each triangle Ω_k,

(ii) $\nu_k \cdot T z_k = -\nu_l \cdot T z_l$ on $\overline{\Omega_k} \cap \overline{\Omega_l}$ when this set is a common side of the two triangles Ω_k, Ω_l,

(iii) $2/r_k = B v_k(x^{(k)})$ for one point $x^{(k)} \in \overline{\Omega_k}$,

(iv) If $\overline{\Omega_k} \cap \overline{\Omega_l}$ is a common side, then $z_k(x^{(k)}) = z_l(x^{(k)})$ for an $x^{(k)} \in \overline{\Omega_k} \cap \overline{\Omega_l}$, for example, for the center of the common side,

(v) $\nu_k \cdot T z_k = \cos \gamma$ on $\partial\Omega \cap \partial\Omega_k$ when this set is a side of Ω_k.

The assumptions (ii)–(v) define a nonlinear system of equations where the unknowns are the boundary contact angles of the spheres with the plane perpendicular to the (x_1, x_2)–plane which contains the common side of two triangles. From equation (7.40), assumptions (i) and (iii) one concludes that

$$|\text{div } Tv_k - Bv_k| \leq \frac{B}{2} \text{ diam}(\Omega_k)$$

in Ω_k.

A capillary surface over a quadrangle was calculated with this method (200 triangles, 280 equations) on a notebook, see [Miersemann (1996)].

Remark 7.14. Probably the first scheme for the calculation of capillary surfaces over general domains was proposed in [Hornung and Mittelmann (1989)]. For schemes with software packages see [Bank (1994); Brakke (1992)].

7.3 Problems

(1) Show that

$$\phi_0(y; \gamma) = \frac{2}{3} \frac{1 - \sin^3 \gamma}{\cos^3 \gamma} - \frac{1}{\cos \gamma} \sqrt{1 - r^2 \cos^2 \gamma},$$

which is defined in the theorem of Sec. 7.2.2, is uniformly bounded with respect to $0 < r < 1$ and $0 \leq \gamma \leq \pi$.

(2) Show that the angle $2\omega_l$, which is defined in Sec. 7.2.5, is the angle at the corner of the capillary surface about the related corner in the base domain.

(3) Conjecture: Proposition 7.5 holds if $\gamma = 0$, provided the boundary of Ω is sufficiently smooth.

(4) Write w_0^*, defined in Sec. 7.2.5 in the example of an annulus, in terms of elliptic integrals.

(5) Write C_0, defined in Sec. 7.2.5 in the example of an annulus, in terms of elliptic integrals and Jacobian elliptic functions.

(6) Show that the infinite series $\sum_{l=0}^{\infty} h_{4l-1}(\theta; \alpha, \gamma) r^{4l-1}$, see Proposition 7.6, can not converge pointwise to $u(x)$.

(7) Show that the asymptotic formula of Proposition 7.6 is uniform with respect to $\gamma \to 0$. More precisely, suppose that γ_0 satisfies $\alpha + \gamma_0 < \pi/2$, then the constants A and r_0 are independent on γ, $0 < \gamma \leq \gamma_0$. *Hint:*

Let $h(\theta; \gamma) := h_{-1}(\theta; \alpha, \gamma)$ and $q(\theta, \gamma) := h_{4l-1}(\theta; \alpha, \gamma)$, $l \geq 1$. Then

$$\sup |q(\theta, \gamma)| < \infty,$$

$$\sup \frac{|q_\theta(\theta, \gamma)|}{(h^2(\theta; \gamma) + h_\theta^2(\theta; \gamma))^{1/2}} < \infty,$$

$$\sup \frac{|q_{\theta\theta}(\theta, \gamma)|}{(h^2(\theta; \gamma) + h_\theta^2(\theta; \gamma))^{3/2}} < \infty,$$

where the supremum is taken over $\theta \in (-\alpha, \alpha)$ and $0 < \gamma \leq \gamma_0$. Then follow the calculations in [Miersemann (1993)].

(8) Consider the problem of two parallel plates with distance d, see Sec. 3.3.2. Show that

$$u = \frac{\cos \gamma_1 + \cos \gamma_2}{\kappa d} + O(d)$$

as $d \to 0$.

(9) Determine the solution of the ascent between two parallel plates, see Sec. 3.3.2, by using a numerical method adapted from Sec. 7.2.7, and discuss the sign of $\mathcal{F}_x^L(d, \gamma_w, \gamma_l)$, see Sec. 5.2.1 (attraction/repelling of a plate).

Chapter 8

Surface tension

8.1 Surface tension

Some methods of measurement of surface tension are based on the equilibrium conditions for floating bodies. Another method exploits the ascent of a liquid in a capillary tube. These methods require the knowledge of the boundary contact angle. A new method is based on the measurement of the ascents as well as at the interior and at the exterior wall of a capillary tube. This method does not require the knowledge of the contact angle provided the contact angles at the interior and the exterior wall of the tube are equal. We recall, see Chap. 1, that the surface tension is defined by

$$\sigma = \frac{\triangle W}{\triangle A} \ [mN/m], \tag{8.1}$$

where $\triangle W$ is the work to get the additional surface $\triangle A$.

8.2 Force methods

Assume that a cylindrical homogeneous body $\Omega = D \times [0, l]$ with constant cross section D hangs vertically in a liquid which fills partly a container, see Fig. 8.1. The force F is directed upwards in order to keep the body in an equilibrium. We suppose that the body can move in the x_3-direction only and that no rotation is possible. Let $z = v(x)$, $x = (x_1, x_2)$, define the free surface $\mathcal{S}$. Then, see the remark below of Theorem 5.1, the equilibrium condition is given by

$$\sigma \int_{\partial \mathcal{S}} \langle \nu, e_3 \rangle \, ds + g\rho_1 \int_{\mathcal{W}(\mathcal{S})} y_3(v)(x) \langle N_\Sigma, e_3 \rangle \, dA + g\rho_2 |\Omega|$$

$$+ \lambda_0 \int_{\mathcal{W}(\mathcal{S})} \langle N_\Sigma, e_3 \rangle \, dA = F,$$

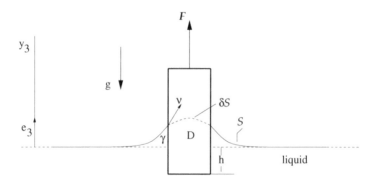

Fig. 8.1   Measurement of surface tension

where

$W(S)$ denotes the wetted part of the boundary Σ of the body Ω,
N_Σ the exterior normal at Σ,
ρ_1 density of the liquid,
ρ_2 density of the body.

The constant λ_0 can be calculated explicitly. It depends on the volume of the liquid and on the location of the body Ω. Set

$$v = u - \frac{\lambda_0}{g\rho_1},$$

then u satisfies the equilibrium conditions

$$\sigma \int_{\partial_2 S} \langle \nu, e_3 \rangle \, ds + g\rho_1 \int_{W(S)} y_3(u)\langle N_\Sigma, e_3 \rangle \, dA + g\rho_2 |\Omega| = F. \qquad (8.2)$$

Here u denotes the ascent of the liquid from the "general level", the dashed line, which is approximately a horizontal plane in the case that the body is sufficiently far away from the boundary of the container and if the gravity is large enough. Above of the (x_1, x_2)-plane the ascent $u(x)$ is close to zero, see estimates derived in [Siegel (1980)]. Numerical calculations for rotationally capillary surfaces suggest that the error is much smaller.

From formula (8.2) we get

$$\sigma |\partial S| \cos \gamma_2 + g\rho_1 h |D| + g\rho_2 |\Omega| = F,$$

where h denotes the distance of the bottom of Ω from the "general line" which is positive if the bottom is above of the general line and negative if the bottom is below of this line. Then

$$\sigma = \frac{F - g\rho_1 h |D| - g\rho_2 |\Omega|}{|\partial S| \cos \gamma}. \qquad (8.3)$$

The force F and the distance h between the bottom of the body and the dashed line are known from measurements. In general, the contact angle γ and $|\partial\mathcal{S}|$ are unknown.

8.2.1 *Wilhelmy plate method*

The body is here a thin plate where the cross section is a thin rectangle, see Fig. 8.2. In the method of [Wilhelmy (1863)], it is assumed that the

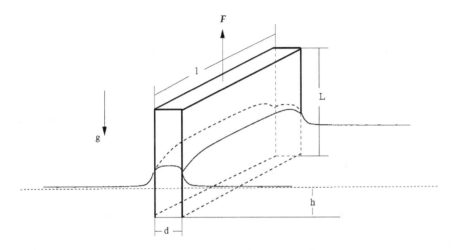

Fig. 8.2 Wilhelmy plate

contact angle γ_2 is zero, and $|\partial\mathcal{S}|$ is replaced by $2l+2d$. Thus, the Wilhelmy method yields an upper bound σ^* for the surface tension σ, where

$$\sigma^* = \frac{F - g\rho_1 hld - g\rho_2|\Omega|}{2l + 2d}.$$

Then we get for a thin plate with $d \approx 0$ that $\sigma \approx F/(2l)$.

8.2.2 *du Noüy ring method*

The method of [Noüy (1925)] uses a ring which will be pulled out of a liquid, see Figs. 8.3 and 8.4. From the equilibrium condition

$$\sigma \int_{\partial\mathcal{S}} \langle \nu, e_3 \rangle \, ds + g\rho_1 \int_{\mathcal{W}(\mathcal{S})} y_3(v)\langle N_\Sigma, e_3 \rangle \, dA + g\rho_2|\Omega| = F$$

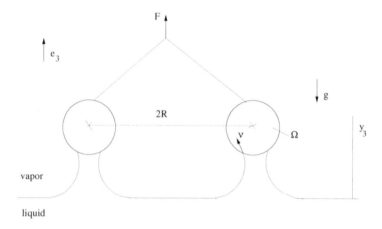

Fig. 8.3   du Noüy ring method in detail

one finds approximately the surface tension when the ring is thin, the force is maximal and the capillary surface consists of two parallel cylinders, see Fig. 8.4. The above assumptions imply that

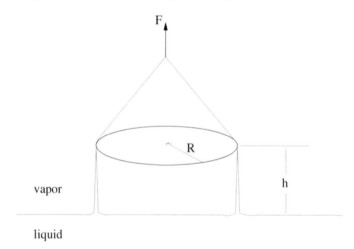

Fig. 8.4   du Noüy ring method

$$\sigma \approx \frac{F}{4\pi R}.$$

The right hand side is exactly the formula (8.1) which defines the surface tension.

8.2.3 *Padday method*

The capillary surface hangs on the lower edge of a rod with constant circular cross section, see Fig. 8.5.

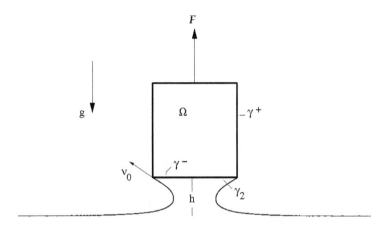

Fig. 8.5   Padday device for measurement of surface tension

From formula (8.2) we see that

$$\sigma = \frac{F - g\rho_1 h\pi R^2 - g\rho_2\pi LR^2}{2\pi R \sin\gamma_2}.$$

The contact angle γ_2 is not determined. Assume that the capillary surface is in an equilibrium, then, see Theorem 6.1, γ_2 satisfies the inequalities

$$\gamma^- \le \gamma_2 \le \gamma^+ + \frac{\pi}{2},$$

where γ^-, γ^+ denote the contact angle associated to the bottom and to the cylindrical surface, resp., of the rod. In [Padday, Pitt and Pashley (1975)] it is proposed to determine the surface tension from the maximum pull on the rod.

8.3   An ascent method

In the following we consider a new method[1] how to find the surface tension *without any assumption on the contact angle*, see [Miersemann (2012)]. This parameter identification method is based on the simultaneous measurement of the ascents $\triangle_e$ and $\triangle_i$ at the exterior and the interior at a circular capillary tube, see Fig. 8.6. This method needs gravity directed downwards.

---

[1]Patent DE 10 2011 009 144, [Finn and Miersemann (2014)]

We consider a transparent tube with a constant circular cross section dipped into the middle of a container which is filled with liquid. For the

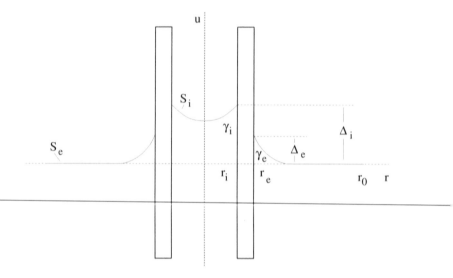

Fig. 8.6　Ascents of liquid at a tube

following notations see Fig. 8.6. Let
r_e exterior radius of the tube,
r_i interior radius of the tube,
γ_e contact angle at the exterior of the tube,
γ_i contact angle at the interior of the tube.

Suppose that the contact angles are constant and are in the closed interval $[0, \pi/2]$. The interior capillary surface S_i above of the general level is defined by $z = u_i(r; \gamma_i, \kappa)$, where u_i is the solution of the Laplace boundary value problem

$$\left(\frac{r u'(r)}{\sqrt{1 + (u'(r))^2}} \right)' = \kappa r\, u(r) \quad \text{in } 0 < r < r_i, \tag{8.4}$$

$$\lim_{r \to r_i - 0} \frac{u'(r)}{\sqrt{1 + (u'(r))^2}} = \cos \gamma_i, \tag{8.5}$$

$$u'(0) = 0. \tag{8.6}$$

The exterior surface S_e over $r_e < |x| < r_0$ is defined by $w_e = w_e(r; \gamma_e, \kappa)$, where w_e is the solution of

$$\left(\frac{rw'(r)}{\sqrt{1+(w'(r))^2}} \right)' = \kappa\, r\, w(r) \quad \text{in} \ \ r_e < r < r_0, \tag{8.7}$$

$$\lim_{r \to r_e+0} \frac{w'(r)}{\sqrt{1+(w'(r))^2}} = -\cos\gamma_e, \tag{8.8}$$

$$w'(r_0) = 0. \tag{8.9}$$

Suppose that r_0, $r_e < r_0$, is sufficiently large. Then $u_i(r_i; \gamma_i, \kappa)$ is the ascent at the interior wall and $w_e(r_e; \gamma_e, \kappa)$ is the ascent at the exterior wall of the tube up to a small error which we can neglect, see [Miersemann (2012)] for details. Let $\triangle_i$ and $\triangle_e$ are the ascents measured from the general level, see Fig. 8.6. Then the equations

$$\triangle_i = u_i(r_i; \gamma_i, \kappa),$$
$$\triangle_e = w_e(r_e; \gamma_e, \kappa)$$

define functions $\kappa i(\gamma_i)$ and $\kappa e(\gamma_e)$. Assume that the *interior and the exterior contact angle are the same*, then we get from equation

$$\kappa i(\gamma) = \kappa e(\gamma)$$

both, the contact angle and the capillarity constant. Here we have used that the surface tensions of the liquids in and outside of the tube are equal since the same liquid is in and outside of the tube.

Asymptotic formulas for the ascent in the interior of the tube, see [Laplace (1806); Siegel (1980)] and for the ascent at the exterior of the tube, see [Lo (1983); Miersemann (2006)], support the following conjectures.

Conjecture 8.1. *The function $\kappa i(\gamma)$ is strictly concave from below and $\kappa e(\gamma)$ is strictly convex from below, see Fig. 8.7.*

Conjecture 8.2. *The curves $\kappa i(\gamma)$ and $\kappa e(\gamma)$ intersects at one point only, see Fig. 8.7.*

The previous conjecture is supported through numerical calculations, see [Miersemann (2012)], p. 4, and by the following inclusion result. Moreover, two points of intersection would imply, caused by the strong monotony of the curves, two different surface tensions for the same liquid.

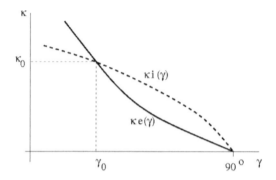

Fig. 8.7   Intersection of the curves $\kappa i(\gamma)$ and $\kappa e(\gamma)$

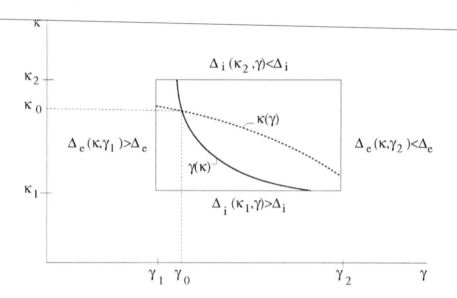

Fig. 8.8   Inclusion, notations

Lemma 8.1. *Suppose that in the intervals* $0 < \kappa_1 < \kappa_2$ *and* $0 < \gamma_1 < \gamma_2 < 90^o$ *the inequalities*

$$\triangle i(\kappa_1, \gamma) > \triangle_i, \quad \triangle i(\kappa_2, \gamma) < \triangle_i, \quad \gamma_1 \leq \gamma \leq \gamma_2,$$
$$\triangle e(\kappa, \gamma_1) > \triangle_e, \quad \triangle e(\kappa, \gamma_2) < \triangle_e, \quad \kappa_1 \leq \kappa \leq \kappa_2,$$

are satisfied. Then there is a solution κ_0, γ_0 *of the system*

$$\triangle i(\kappa, \gamma) = \triangle_i,$$
$$\triangle e(\kappa, \gamma) = \triangle_e$$

in the rectangle $\kappa_1 < \kappa < \kappa_2, \ \gamma_1 < \gamma < \gamma_2.$

Proof. The assumptions imply the existence of a function $\gamma(\kappa)$ defined on $\kappa_1 \leq \kappa \leq \kappa_2$ where $\gamma_1 < \gamma(\kappa) < \gamma_2$, and of a function $\kappa(\gamma)$ defined on $\gamma_1 \leq \gamma \leq \gamma_2$ with $\kappa_1 < \kappa(\gamma) < \kappa_2$, see. Fig. 8.8. The comparison principle implies that the functions $\kappa(\gamma)$ and $\gamma(\kappa)$ are strongly decreasing with growing contact angle γ or growing capillarity constant κ. □

We solve the above boundary value problems (8.4)–(8.6) and (8.7)–(8.9) numerically with a method proposed in [Miersemann (1996)], see Sec. 6.2.7. This procedure is a finite element method where the elements are capillary surfaces between narrow coaxial circular cylinders. This method admits an explicit error estimate of order $O(\sqrt{h})$ uniformly in $\gamma_i, \ \gamma_e, \ \gamma_R \in [0, \pi]$, provided that $\kappa h^2 \leq c$ with a known constant c. Here h denotes the step size of the method.

Remark 8.1. Asymptotic formulas for the ascent in the interior of the tube, see [Laplace (1806); Siegel (1980)] and, in particular, for the ascent at the exterior of the tube, see [Lo (1983); Miersemann (2006)] are too rough and yield unrealistic results for the surface tension.

Experiments with a glass tube show that we get realistic results, see [Miersemann (2012)]. In the case of distilled water of 20° Celsius we obtained that $\kappa = g\rho/\sigma \approx 13.3 \, cm^{-2}$, and for water with some surfactant $\kappa \approx 35 \, cm^{-2}$.

Remark 8.2. Once one knows the surface tension then we find the contact angle on a wire or on a rod with constant circular cross section from the ascent of liquid at the exterior of the related cylinder, see [Miersemann (2012)], p. 5, for an example.

8.4 Problems

(1) Write a code which yields the solution θ (and then κ) of $\kappa_i(\theta) = \kappa_e(\theta)$.

(2) Carry out experiments with a glass tube of large exterior and small interior radius, and with different kinds of liquids.

(3) Discuss the above conjectures.

Chapter 9

Contact angle

According to the classical theory of [Gauss (1839)], the boundary contact angle of the capillary surface with a homogeneous container wall is constant. On the other hand it was discovered by [Shinbrot (1985)] that a drop on an inclined plane has a non-constant contact angle along its trace on the plane, provided the field of gravity is not perpendicular to the plane. In order to get a theory in agreement with reality [Finn and Shinbrot (1988)], Chap. 8.1, made the assumption of an additional resistance force. If one considers the boundary of the wetted part as a wetting barrier then we can solve probably the conflict between the reality and the theory of Gauss.

9.1 Liquid on a homogeneous container wall

In the following we assume that the adhesion coefficients of the wetted part and the non-wetted part of the wall are different in general. Then we get inequalities for the contact angle.

As in Chap. 3 we define a family of admissible comparison surfaces $\mathcal{S}^\pm(\epsilon)$ by

$$z^\pm(u, \epsilon) = x(u) + \epsilon \zeta^\pm(u) + O(\epsilon^2),$$

where $0 < \epsilon < \epsilon_0$, and

$$\zeta = \zeta^\pm = \xi^\pm(u) N_{\mathcal{S}_0}(u) + \eta^\pm(u) T_{\mathcal{S}_0}(u)$$

define comparison surfaces $\mathcal{S}^\pm(\epsilon)$ where the boundary of $\mathcal{S}^+(\epsilon)$ is outside of the wetted part of the container, see Fig. 9.1, and the boundary of $\mathcal{S}^-(\epsilon)$ is in the wetted part. Let β_{nw} be the adhesion coefficient of the non-wetted part, then we get as in the proof of Theorem 3.1 the inequality

$$\int_{\partial \mathcal{S}_0} [-\beta_{nw} \xi^+ \sin \gamma + \eta^+ (1 - \beta_{nw} \cos \gamma)] \, ds \geq 0.$$

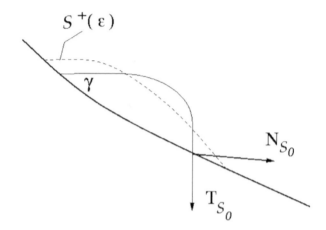

Fig. 9.1   Comparison surface

Set $\xi^+ = \tau \, \sin \gamma$, $\eta^+ = \tau \, \cos \gamma$, $\tau > 0$. Then we find the inequality $\beta_{nw} \leq \cos \gamma$. Analogously, we define $\mathcal{S}^-(\epsilon)$ through $\xi^- = -\tau \, \sin \gamma$ and $\eta^- = -\tau \, \cos \tau$, $\tau > 0$. This leads to the inequality $\beta_w \geq \cos \gamma$. Summarizing, we have

Theorem 9.1. *Suppose β_w and β_{nw} are the adhesion coefficients of the wetted part and the non-wetted part of the container wall. Then the boundary contact angle, not necessarily constant, satisfies the inequalities*

$$\beta_{nw} \leq \cos \gamma \leq \beta_w.$$

As a consequence we get the expected inequality

$$\beta_{nw} \leq \beta_n.$$

The previous variational argument leads to the

Hypothesis 9.1. The adhesion coefficient β_w on the wetted part and the adhesion coefficient β_{nw} on the non-wetted part of a container wall are in general different from each other.

Remark 9.1. Probably, the above theorem leads to a correct definition of the *advancing contact angle* γ_{nw} and the *receding contact angle* γ_w by setting $\cos \gamma_{nw} = \beta_{nw}$ and $\cos \gamma_w = \beta_w$.

For more remarks and hints to some literature concerning the disagreement between experiments and the theory see [Finn (1986)], Chap. 8.1.

9.2 Liquid on surfaces with a definite structure

The contact angle is influenced through the structure of the surface of the container wall. This and related problems are addressed for example in [Schultze (1924, 1925); Quéré (2005)] and [Adamson and Gast (1997)], Chap. X-5.B. We will consider two explicit surface structures which are probably of interest for producing of fabrics. Fig. 1.19 of the introduction shows drops on different materials. Both surfaces are coated with the same wax spray, but the contact angle on the fabric is bigger than the contact angle of the drop on the foil.

The following formal calculation indicate how the structure of the surface defines different contact angles. Assume that the drop sits on a structure with holes, see Fig. 9.2, and that the pressure under the fabric is the same as above of the material. In Fig. 9.2 the contact angle between the

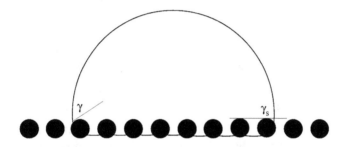

Fig. 9.2   Structured surface

threads of the material is γ and γ_s is the angle between the flat surface and the drop.

In the following we assume that the drops are small. Then we can neglect the gravity since drops of small volume are close to drops in a zero gravitational field, see [Finn (1986)], Chap. 3, and [Miersemann (1994)].

The energy E of the drop is, see Chap. 3.1, given by

$$E = \sigma \left(|S| - \beta |S^*| \right) ,$$

where σ is the surface tension, β, $-1 \le \beta \le 1$, the local adhesion coefficient, $|S|$ the area of the interface and $|S^*|$ is the area of the wetted surface. If the drop is in an equilibrium state, then $\beta = \cos \gamma$, see Chap. 3.1. Thus

$$E = \sigma \left(|S| - |S^*| \cos \gamma \right)$$

is the energy of the drop in an equilibrium.

9.2.1 *A model*

We assume that the liquid is a cap sitting on a structured surface, see Fig. 9.3.

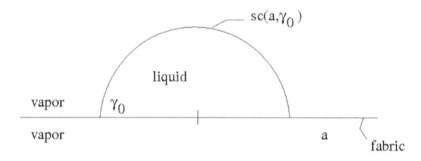

Fig. 9.3   Drop on a structured surface

Let γ be the contact angle between the liquid and the material from which the surface is made, see Fig. 9.2. Consider a domain Ω of the macroscopic surface on which the liquid sits. Then we assume that the non-wetted and the wetted domains n, w, resp., are given by

$$n(\Omega, \gamma) = |\Omega| n_0(\gamma) , \quad w(\Omega, \gamma) = |\Omega| w_0(\gamma),$$

where $|\Omega|$ denotes the area of Ω, and n_0, w_0 are non-negative functions depending only on γ. Then the energy of the drop in equilibrium on the structure, see Fig. 9.3, is defined through

$$E = \sigma \left(|S| - |S^*| \cos \gamma \right) ,$$

where

$$|S| = |sc(a, \gamma_0)| + a^2 \pi n_0(\gamma) \text{ und } |S^*| = a^2 \pi w_0(\gamma) .$$

Here $|sc(a, \gamma_0)|$ denotes the area of the cap $sc(a, \gamma_0)$, see Fig. 9.3.

Let $|V|$ be the volume of the drop $V(a, \gamma_0)$, $0 < a < \infty$, and set

$$f(a, \gamma_0, \gamma) := |sc(a, \gamma_0)| + a^2 \pi n_0(\gamma) - a^2 \pi w_0(\gamma) \cos \gamma .$$

Then $\sigma f(a, \gamma_0, \gamma)$ is the energy of the drop,

Definition 9.1. The angle $\gamma_0 \in (0, \pi)$ is called *macroscopic contact angle* if there is an $\epsilon_0 > 0$ such that

$$f(a + \epsilon, \gamma_0 + \delta(\epsilon), \epsilon) \geq f(a, \gamma_0, \gamma) \tag{9.1}$$

for all ϵ, $|\epsilon| < \epsilon_0$. Here is $\delta = \delta(\epsilon)$ the solution of the equation

$$|V(a, \gamma_0)| = |V(a + \epsilon, \gamma_0 + \delta)| .$$

Inequality (9.1) means that the energy of a drop with the macroscopic contact angle γ_0 defines a local minimum in comparison with other drops with the same volume. According to a result of [Gonzales (1976)] it is sufficient to compare with caps only.

From inequality (9.1) we obtain the equation

$$\cos \gamma_0 = w_0(\gamma) \cos \gamma - n_0(\gamma) , \qquad (9.2)$$

which defines formally the macroscopic contact angle $\cos \gamma_0$, provided that

$$-1 < w_0(\gamma) \cos \gamma - n_0(\gamma) < 1 .$$

holds.

In the following two examples we calculate approximately the functions $n_0(\gamma)$ and $w_0(\gamma)$. The resulting macroscopic contact angles are qualitatively in agreement with the observed angles.

9.2.2 *Fabrics, sieves*

Consider a net made from threads of circular cross section with a constant radius r as shown in Fig. 9.4.

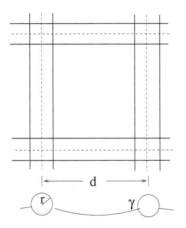

Fig. 9.4   Fabric or sieve structure

Let d be the distance between the middle of two parallel threads. Assume that $r = qd/2$ for a fixed positive constant q, $0 < q < 1$. An elementary calculation shows that we get approximately

$$w_0(\gamma) = 2(\pi - \gamma)q + O(q^2),$$
$$n_0(\gamma) = (1 - q \sin \gamma)^2 + O(q^2)$$
$$= 1 - 2q \sin \gamma + O(q^2) .$$

Example 9.1. Setting $d = 1.6$ mm and $r = 0.15$ mm, then $q = 0.18$. If $\gamma = 80^0$, then one finds from (9.2) for the macroscopic contact angle $\gamma_o \approx 123^0$, provided that the O-terms are neglected. In experiments with a sieve one observes macroscopic contact angles bigger than 90^0.

9.2.3 Knops structure

We assume that the knops are small balls with radius r and the centers are located in a plane as sketched in Fig. 9.5.

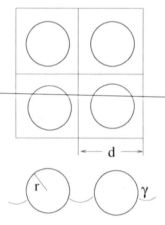

Fig. 9.5   Knops structure

Set $r = qd/2$ for a fixed constant q, $0 < q < 1$. An elementary calculation supports the assumption that

$$w_0(\gamma) = \frac{1}{2}q^2\pi(1 + \cos\gamma) + O(q^3),$$

$$n_0(\gamma) = 1 - \frac{1}{4}q^2\pi\sin^2\gamma + O(q^3) .$$

Example 9.2. Set $q = 0.1$. If $\gamma = 80^0$, then one finds from (9.2) for the macroscopic contact angle $\gamma_o \approx 166^0$, provided that the O-terms are neglected.

9.2.4 Drop on a ring

Here a small drop sitting on a single ring is considered, see Fig. 1.8. If the drop is of small volume then the gravity can be neglected. The question which we ask in the following is whether or not the drop can rest on the ring. The drop is said to be (weakly) stable if the second variation of the associated energy functional is positive for all non-negative and admissible variations. Assume that the drop is a part of a ball with radius R and is sitting on a ring (torus). Let r be the radius of the cross section of the torus, see Fig. 9.6.

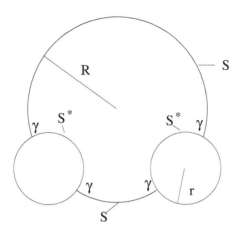

Fig. 9.6   Notations, drop on a ring

The surface of the drop S has the constant mean curvature $1/R$, and S meets the torus in the constant contact angle γ, $0 < \gamma < \pi$. Such a drop defines an equilibrium configuration since the necessary equilibrium conditions of Theorem 3.1 with $F = 0$ are satisfied.

9.2.5   A stability criterion

A necessary criterion for the stability of the drop is that the second variation of the energy functional is positive for all non-negative admissible variations. In the case of a drop on a torus the criterion reads as

$$\int_S \left(|\nabla \zeta|^2 - \frac{2}{R^2} \zeta^2 \right) dA + \int_{\partial S} \left(-\frac{1}{R} \frac{\cos \gamma}{\sin \gamma} + \frac{1}{r} \frac{1}{\sin \gamma} \right) \zeta^2 \, ds > 0 \quad (9.3)$$

for all $\zeta \in W^{1,2}(S)$, $\zeta \neq 0$ and $\int_S \zeta \, dA = 0$. Here ∇ denotes the gradient operator on the surface S and ∂S is the boundary of S. Concerning the definition of ∇ see for instance [Blaschke (1921)]. The second variation was firstly calculated by [Wente (1966)], see also [Wente (1980)] and the corollary to Lemma 3.1.

9.2.5.1 *An equivalent eigenvalue criterion*

Inequality (9.3) is equivalent to an inequality for an associated eigenvalue problem

Set

$$a(\zeta, \phi) := \int_S \left(\nabla \zeta \nabla \phi + \frac{1}{R^2} \zeta \phi \right) dA$$

$$b(\zeta, \phi) := \frac{1}{R^2} \int_S 3 \zeta \phi \, dA - h(r, \gamma) \int_{\partial S} \zeta \phi \, ds \, ,$$

where

$$h(r, \gamma) := -\frac{1}{R} \frac{\cos \gamma}{\sin \gamma} + \frac{1}{r} \frac{1}{\sin \gamma}$$

and let

$$V := \{ \phi \in W^{1,2}(S); \int_S \phi \, dA = 0 \} \, .$$

We will see that inequality (9.3) holds if and only if the lowest positive eigenvalue λ_1^+ of the problem

$$\zeta \in V \setminus \{0\} : \quad a(\zeta, \phi) = \lambda b(\zeta, \phi) \text{ for all } \phi \in V \tag{9.4}$$

satisfies the inequality $\lambda_1^+ > 1$. If $\lambda_1^+ < 1$, then one can show that there are admissible variations which lead to smaller value of the associated energy. That is, the drop is not stable on the ring, it slips probably through the ring.

The equivalence of the inequalities (9.3) and $\lambda_1^+ > 1$ is a consequence of the following reasoning. The quadratic form $a(\phi, \phi)$ is equivalent to a norm on V and $b(\phi, \phi)$ is a completely continuous form on V. Consequently, the spectrum of the eigenvalue problem (9.4) consists of countable many positive and negative eigenvalues. The lowest positive eigenvalue λ_1^+ is defined by

$$\left(\lambda_1^+ \right)^{-1} = \max_{\phi \in V \setminus \{0\}} \frac{b(\phi, \phi)}{a(\phi, \phi)} \, .$$

The definition of $b(\phi, \phi)$ implies that the right hand side of the previous equation is positive. From the definition of λ_1^+ we see that

$$a(\phi, \phi) - b(\phi, \phi) \geq \left(1 - \left(\lambda_1^+ \right)^{-1} \right) a(\phi, \phi)$$

for all $\phi \in V$. Consequently, inequality (9.3) holds if $\lambda_1^+ > 1$.

Assume now that inequality (9.3) is satisfied. Consider the maximum problem $\max_{\phi \in V} b(\phi, \phi)$ under the side condition $a(\phi, \phi) \leq 1$. There exists a solution ϕ_0 since $a(\phi, \phi)$ is equivalent to a norm and $b(\phi, \phi)$ is completely continuous on V. We have that $a(\phi_0, \phi_0) = 1$, since the assumption $a(\phi_0, \phi_0) < 1$ leads to a contradiction as follows. There is a $\phi \in V$ satisfying $b(\phi, \phi) > 0$. Consequently, $b(\phi_0, \phi_0) > 0$ and there is a $t > 1$, such that $t\phi_0$ satisfies the side conditions. From $b(t\phi_0, t\phi_0) = t^2 b(\phi_0, \phi_0) > b(\phi_0, \phi_0)$ we get a contradiction to the fact that ϕ_0 solves the above maximum problem.

Then λ_1^+ is given through $(\lambda_1^+)^{-1} = b(\phi_0, \phi_0)$. The inequality $\lambda_1^+ > 1$ follows since $a(\zeta, \zeta) - b(\zeta, \zeta)$ is the left hand side of the inequality (9.3) and that $a(\phi_0, \phi_0) - b(\phi_0, \phi_0) = 1 - (\lambda_1^+)^{-1}$ holds.

The idea to solve a variational problem in a convex set and show then that the solution is on the boundary of this set one finds in [Beckert (1968)].

9.2.6 *Problems*

(1) Discuss the above Hypothesis 9.1, probably by using experiments.
(2) Show that inequality (9.1) implies equation (9.2).
(3) Discuss the above problem of the drop on a ring numerically.

Chapter 10

Lagrange multiplier rules

There is a large variety of Lagrange multiplier rules for equations and in-equalities. We have used two of them in previous sections. For the convenience of the reader we will state and prove the rules related to our problems. The following Lemma 10.1 and Lemma 10.2 can be easily extended to more than one side condition.

Let H be a real Hilbert space with the inner product $\langle u, v \rangle$, where $u,\ v \in H$.

10.1 Equations

Let $V \subset H$ be a non-empty subspace. For example, in the case of a floating smooth particle the necessary conditions (5.7) and (5.8) define an equation $f'(x)(h) = 0$ for all $h \in V \cap Z$, where $f'(x)$ is a bounded linear functional on H, $Z = \{h \in V : \ g'(x)(h) = 0\}$. Here $g'(x)$ stands for another bounded linear functional defined on H.

Lemma 10.1. *The previous assumptions imply the existence of a real parameter λ_0 such that*

$$f'(x)(w) + \lambda_0\, g'(x)(w) = 0$$

for all $w \in V$ holds.

Proof. There are $F,\ G \in \operatorname{cl} V$, where $\operatorname{cl} V$ denotes the closure of V with respect to the Hilbert space norm, such that

$$f'(x)(h) = \langle F, h \rangle, \quad g'(x)(h) = \langle G, h \rangle$$

for all $h \in \operatorname{cl} V$. Set $Y = \operatorname{span} G$, then $\operatorname{cl} V = Y \oplus Y^{\perp}$. and $F = F_1 + F_2$, where $F_1 \in Y$ and $F_2 \in Y^{\perp}$. Since $\langle F, F_2 \rangle = 0$, we get $F_2 = 0$. Consequently, $F + \lambda_0\, G = 0$, or

$$\langle F, h \rangle + \lambda_0\, \langle G, h \rangle = 0$$

for all $h \in$ cl V. $\qquad\qquad\qquad\qquad\qquad\qquad\qquad\qquad\qquad\quad$ □

10.2 Inequalities

In the case of containers or particles with wetting barriers we are led to $f'(x)(h) \geq 0$ for all $h \in K \cap Z$, where $K \subset V$ is a nonempty convex cone with vertex at zero, and $Z = \{h \in V : g'(x)(h) = 0\}$. We recall that K is said to be a cone with vertex at zero if $h \in K$ implies that $t\, h \in K$ for all $t > 0$. By C^* we denote the polar cone of a cone with vertex at the origin. We recall that the polar cone of a cone $C \subset$ cl V with the vertex at zero is defined to be the cone $C^* = \{v \in$ cl $V : \langle v, w \rangle \leq 0$ for all $w \in C\}$, see [Rockafellar and Wets (1998)], p. 215, for example.

Lemma 10.2. *Suppose that there is an $h_0 \in K$ such that $-h_0 \in K$ and $g'(x)(h_0) \neq 0$. Then there exists a real λ_0 such that*

$$f'(x)(w) + \lambda_0\, g'(x)(w) \geq 0$$

for all $w \in K$.

Proof. Following the proof of the previous lemma, we find that $\langle F, h \rangle \geq 0$ for all $h \in$ cl $K \cap$ cl Z, i. e., $-F \in ($cl $K \cap$ cl $Z)^*$. Then the proof of the lemma is based on the formula, see Lemma 10.3 below,

$$(\text{cl } K \cap \text{cl } Z)^* = \text{cl } (K^* + Z^*).$$

Thus, since $Z^* = \text{span } \{G\}$, it follows

$$-F \in \text{cl } (K^* + \text{span } \{G\}).$$

Consequently, there are sequences $z_n \in K^*$, $y_n \in$ span $\{G\}$ such that $z_n + y_n \to -F$ in cl V. If the sequence y_n remains bounded, then there is a convergent subsequence $y_{n'} \to y$, and $z_{n'} \to z \in K^*$ which implies that $-F \in K^* + y$. Thus there is a real λ_0 satisfying $-F - \lambda_0 G \in K^*$, or equivalently, $\langle F + \lambda_0 G, h \rangle \geq 0$ for all $h \in$ cl K.

Suppose that the sequence $y_n \in$ span $\{G\}$ is unbounded. Set $w_n = z_n + y_n$, then $w_n - y_n = z_n \in K^*$. Thus $\langle w_n - y_n, h \rangle \leq 0$ for all $h \in$ cl K, or

$$\langle w_n, h \rangle - \lambda_n \langle G, h \rangle \leq 0$$

for all $h \in$ cl K. Since $|\lambda_n| \to \infty$, we get $\langle G, h \rangle \leq 0$ for all $h \in$ cl K or $\langle G, h \rangle \geq 0$ for all $h \in$ cl K, which is a contradiction to the assumption of the lemma. $\qquad\qquad\qquad\qquad\qquad\qquad\qquad\qquad\qquad\qquad\qquad$ □

Remark 10.1. Applying Lemma 10.2 to the problem considered in Chap. 6, we are led to two variational inequalities

$$f'(x)(w) + \lambda_0^- \, g'(x)(w) \geq 0 \quad \text{for all } h \in X^-,$$
$$f'(x)(w) + \lambda_0^+ \, g'(x)(w) \geq 0 \quad \text{for all } h \in X^+.$$

It follows that $\lambda_0^- = \lambda_0^+$, provided there exists an $h_0 \in X^- \cap X^+$ such that $-h_0 \in X^- \cap X^+$ and $g'(x)(h_0) \neq 0$.

Extending a result of [Rockafellar and Wets (1998)], Corollary 11.25(b), p. 495, or [Rockafellar (1970)], Corollary 16.4.2, p. 146, to a real Hilbert space, we get

Lemma 10.3. *Let H be a real Hilbert space. Suppose that $K_1, \ldots, K_m \subset H$ are nonempty, closed and convex cones with their vertices at the origin. Then*

$$(K_1 \cap \cdots \cap K_m)^* = cl \; (K_1^* \cdots + K_m^*) \; .$$

Proof. (i) The inclusion

$$(K_1^* \cdots + K_m^*) \subset (K_1 \cap \cdots \cap K_m)^*$$

follows since we have for given $v_i \in K_i^*$ that $\langle v_i, h \rangle \leq 0$ for all $h \in K_i$. Consequently, $\langle v_1 + \cdots + v_m, h \rangle \leq 0$ for all $h \in K_1 \cap \cdots \cap K_m$. Thus $v_1 + \cdots + v_m \in (K_1 \cap \cdots \cap K_m)^*$.
(ii) Set $C = cl \; (K_1^* \cdots + K_m^*)$. Let $w \in (K_1 \cap \cdots \cap K_m)^*$ be given and suppose that $w \notin C$. From a separation theorem it follows that there is a $p \in H$ such that $\langle p, w \rangle > 0$ and $\langle p, y \rangle \leq 0$ for all $y \in C$. We have $\langle w, v \rangle \leq 0$ for all $v \in K_1 \cap \cdots \cap K_m$ and $\langle p, y \rangle \leq 0$ for all $y \in K_1^* \cdots + K_m^*$. The previous inequality shows that $p \in K_i$ for all i. Then the inequality $\langle w, p \rangle \leq 0$ is in conflict with the separation theorem. □

10.3 Problems

(1) *Separation theorem.* Let H be a real Hilbert space and V a nonempty, closed and convex subset. Let $w \in H$ and $w \notin V$. Show that there is a real λ such that $\langle p, y \rangle \leq \lambda < \langle p, w \rangle$ for all $y \in V$.
   Hint: Consider the minimum problem $\min_{y \in V} \|y - v\|^2$ and use the Banach-Saks theorem that a closed convex subset is weakly closed.
(2) *Separation theorem.* Let V in the previous exercise be a closed convex cone C with its vertex at zero. Show that $\langle p, y \rangle \leq 0 < \langle p, w \rangle$ for all $y \in C$.

(3) Extend the Lagrange multiplier rule of Lemma 10.1 to more equations as side conditions.

(4) Extend the Lagrange multiplier rule of Lemma 10.2 to more equations as side conditions.

Acknowledgments

First of all I would like to thank Robert Finn for many inspiring conversations over many years. Moreover, I thank Stefan Ackermann for his kindly and permanent help to produce source files for my articles. Let me thank also the Technical Department of the Faculty of Physics and Harald Krautscheid from the Chemistry Department for their support concerning some experiments. Last but not least, I would like to thank Lim Swee Cheng from the World Scientific Publishing house for his encouragement to write this book.

Bibliography

Abramowitz, M. and Stegun, I. A. (1964). *Handbook of Mathematical Functions with Formulas, Graphs, and Mathematical tables*, (National Bureau of Standards Applied Mathematics Series, U.S. Government Printing Office, Washington, DC), Vol. 55, reprinted by Dover, New York, 1972.

Adams R. A. (1975). *Sobolev Spaces*, (Academic Press).

Adamson, A. W. and Gast, A. P. (1997). *Physical Chemistry of Surfaces*, (John Wiley).

Aspley, A., He C. and McCuan, J. (2015). Force profiles for parallel plates partially immersed in a liquid bath, *J. Math. Fluid Mech.* **17**, pp. 87–102.

Bank, R. E. (1994). *PLTMG: A Software Package for Solving Elliptic Partial Differential Equations*, Frontiers in Applied Mathematics, Vol. 15, (SIAM, Philadelphia).

Beckert, H. (1968). Über eine bemerkenswerte Klasse gemischtfreier Variationsprobleme höherer Ordnung, *Math. Zeitschr.* **103**, pp. 18-29.

Bernstein, S. (1927). Sur un théorème de géométrie et son application aux dérivées partielles du type elliptique, *Comm. Soc. Math. de Kharkov (2)* **15**,(1915–1917), pp. 38–45. German translation: Math. Z. **26**, pp. 551–558.

Bieker, S. and Dietrich, S. (1998). Wetting on curved surfaces, *Physica* **A 252**, pp. 85–137.

Blaschke, W. (1921). *Vorlesungen über Differentialgeometrie I.*, (Springer, Grundlehren **1**).

Brakke, K. A. (1992). The surface evolver, *Experimental Mathematics* **1**, pp. 141–165.

Byrd, P. F. and Friedman, M. D. (1971). *Handbook of Elliptic Integrals for Engineers and Scientists*, (Springer, Grundlehren **67**).

Chen, J. T. (1980). On the existence of capillary free surfaces in the absence of gravity, *Pacific J. Math.* **88**, pp. 323–361.

Chen, J. T., Finn, R. and Miersemann, E. (2009). Capillary surfaces in wedge domains, *Pacific J. Math.* **236**, pp. 101–123.

Concus, P. (1968). Static menisci in a vertical right circular cylinder, *J. Fluid Mech.* **34**, pp. 481–485.

Concus, P. and Finn, R. (1970). On a class of capillary surfaces, *J. Analyse Math.*

23, pp. 65–70.

Concus, P. and Finn, R. (1974). On capillary free surfaces in a gravitational field, *Acta Math.* **132**, pp. 177–198.

Concus, P. and Pereyra, V. (1983). Calculating axisymmetric menisci, *Acta Cient. Venezolana* **34**, pp. 89–100.

Derjaguin, B. V. (1939). A theory of interaction of particles in presence of electric double layers and the stability of lyophole colloids and disperse systems, *Acta Physicochim. URSS* **10**, p. 333.

Derjaguin, B. V. (1940). On the repulsive forces between charged colloid particles and on the theory of slow coagulation and stability of lyophobe sols, *Trans. Faraday Soc.* **36**, pp. 203–211.

Derjaguin, B. V. (1946). Theory of the distortion of a plane surface of a liquid by small objects and its application to the measurement of the contact angle of thin filaments and fibres, *Dokl. Akad. Nauk SSSR* **51**, pp. 519–522.

Dierkes, U, Hildebrandt, S, Küster, A and Wohlrab, O (1992) *Minimal Surfaces I*, (Springer, Grundlehren **295**).

Dierkes, U, Hildebrandt, S, and Sauvigny, F. (2010) *Minimal Surfaces*, (Springer, Grundlehren **339**).

Elcrat, A. Neel, R. and Siegel S. (2004). Equilibrium configurations for a floating drop, *J. Math. Fluid Mech.* **6**, pp. 405–429.

Emmer, M. (1973) Esistenza, unicità e regolarità nelle superfici di equilibrio nei capillari, *Ann. Univ. Ferrara Sez. VII* **18**, pp. 79–84.

Evans, L. C. and Gariepy, R. F. (1992). *Measure Theory and Fine Properties of Functions*, Studies in Advanced Mathematics, (CRC Press, Boca Raton).

Finn R. (1986), *Equilibrium Capillary Surfaces*, (Springer, Grundlehren **284**).

Finn, R. (1997). Editor's note to the article of L. Zhou: On stability of a catenoidal liquid bridge, *Pacific J. Math.* **178**, pp. 197-198.

Finn, R. (1999). Capillary surface interfaces, *Notices of the AMS* **46**, pp. 770–781.

Finn, R. (2011). Criteria for floating I, *J. Math. Fluid Mech.* **13**, pp. 103–115.

Finn, R. (2010). On Young's paradox, and the attractions of immersed parallel plates, *Phys. Fluids* **22**), p. 017103.

Finn, R. and Gerhardt, G. (1977). The internal sphere condition and the capillary problem, *Ann. Mat. Pura Appl.* **112**, pp. 13–31.

Finn, R. and Hwang, J. F. (1989). On the comparison principle for capillary surfaces, *J. Fac. Sci. Univ. Tokyo Sect. IA* **36:1**, pp. 131–134.

Finn, R. (2006). The contact angle in capillarity, *Physics of Fluids* **18**, p. 047102.

Finn, R. (2013). Capillary forces on partially immersed plates, *Differential and Difference Equations with Applications*, Pinelas et al. (eds.), (Springer Proceedings in Mathematics and Statistics **47**), pp. 13–25.

Finn, R., McCuan, J. and Wente, H. C. (2012). Thomas Young's surface diagram: its history, legacy, and irreconcilabilities, *J. Math. Fluid Mech.* **14**), pp. 445–453.

Finn, R. and Miersemann, E. (2014). Verfahren zur Messung der Oberflächenspannung von Flüssigkeiten und zur Messung von Kontaktwinkeln, *Patent DE 10 2011 009 1440*.

Finn, R. and Shinbrot, M. (1988). The capillary contact angle. II: the inclined

Ros, A. and Souam, R. (1997). On stability of capillary surfaces in a ball, *Pacific J. Math.* **178**, pp. 345–361.

Runge, C. (1895). Über die numerische Auflösung von Differentialgleichungen. *Math. Annalen* **46**, pp. 167–178.

Scheeffer, L. (1886). Über die Bedeutung der Begriffe "Maximum und Minimum" in der Variationsrechnung, *Math. Annalen* **26**, pp. 197-208.

Schiller, P., Wahab, M. and Mögel, H.-J. (2004). Adhesion force of a wedge, *Langmuir* **20**, pp. 2227–2232.

Schiller, P., Mögel, H.J. and Miersemann, E. (2010). Stability of liquid layers on solid materials, *J. Math. Fluid Mech.* **12**, pp. 293–305.

Scholz, M. (2003). On the asymptotic behaviour of capillary surfaces in cusps, *Z. angew. Math. Phys.* **55**, pp. 216–234.

Scholz, M. (2004). An attempt to explain the ascent of sap in defoliated trees, *J. Math. Fluid Mech.* **6**, pp. 295–310.

Schultze, K. (1924). Über kapillare Erscheinungen, *Koloid-Zeitschrift* **35**, pp. 76-86.

Schultze, K. (1925). Kapillarität, Verdunstung und Auswitterung, *Koloid-Zeitschrift* **36**, pp. 65-78.

Schwarz, H. A. (1890). Ueber ein die Flächen kleinsten Flächeninhalts betreffendes Problem der Variationsrechnung, *Gesammelte Mathematische Abhandlungen, Band I*, pp. 315–362, (Springer, Berlin).

Shinbrot, M. (1985). A remark on the capillay contact angle, *Math. Meth in the Appl. Sci.* **7**, pp. 383–384.

Siegel, D. (1980). Height estimates for capillary surfaces, *Pacific J. Math.* **88**, pp. 471–516.

Siegel, D. (1987). The behavior of a capillary surface for small Bond number, *Variational Methods for Free Surface Interfaces*, pp. 109–113, eds. Concus and R. Finn, (Springer, Berlin).

Slobozhanin, L. A. (1986). Equilibrium and stability of three capillary fluids, *Fluid Dynamics* **21**, pp. 481–485. Translated from Izvestiya Akademii Nauk SSSR, Mekhaika Zhidkosti i Gaza, No. 3, pp. 170–173, May–June, 1986. Original article submitted April 29, 1985.

Tam, L. F. (1987). On the uniqueness of capillary surfaces, *Variational Methods for Free Surface Interfaces*, pp. 99–108, eds. P. Concus and R. Finn, (Springer, Berlin).

Tam, L. F. (1986). The behavior of capillary surfaces as gravity tends to zero, *Comm. Partial Differential Equations* **11**, pp. 851–901.

Taylor, B. (1712). Concerning the ascent of water between two glass planes, *Philos. Trans. Roy. Soc. London* **27**, p. 538.

Turkington, B. (1980). Height estimates for exterior problems of capillary type, *Pacific J. Math.* **88**, pp. 517–540.

Vogel, T. I. (2000). Sufficient conditions for capillary surfaces to be energy minima, *Pacific J. Math.* **194**, pp. 469–489.

Watson, G. N. (1952). *A treatise on the Theory of Bessel Functions*, (Cambridge).

Wente, H. C. (1966). *Existence theorems for surfaces of constant mean curvature and perturbations of a liquid globule in equilibrium*, Ph.D. thesis, Harvard

University, Cambridge, MA, 1966.

Wente, H. C. (1980). The stability of the axially symmetric pendent drop, *Pacific J. Math.* **88**, pp. 421–470.

Wente, H. C. (1999). Stability analysis for exotic containers, *Dynamics of Continuous, Discrete and Impulse Systems* **5**, pp. 151–158.

Wente, H. C. (1998). *Private communication* (July, 1998).

Wente, H. C. (2011). Exotic capillary tubes, *J. Math. Fluid Mech.* **13**, pp. 355–370.

Wilhelmy, L. (1863). Ueber die Abhängigkeit der Capillaritäts-Constanten des Alkohols von Substanz und Gestalt des benetzten festen Körpers, *Ann. Physik* **119**, (6), pp. 177-217.

Young, T. (1805). An essay on the cohesion of fluids, *Philos. Trans. Roy. Soc. London* **95**, pp. 65-87.

Index

CPSIA information can be obtained
at www.ICGtesting.com
Printed in the USA
BVHW040923160320
575132BV00002B/6